50 PROJECTS USING
I.C. CA3130

ALSO BY THE SAME AUTHOR

No. 222	Solid State Short Wave Receivers for Beginners
No. 223	50 Projects Using IC CA3130
No. 224	50 CMOS IC Projects
No. 226	How to Build Advanced Short Wave Receivers
No. 227	Beginners Guide to Building Electronic Projects
No. BP45	Projects in Opto-Electronics
No. BP49	Popular Electronic Projects
No. BP56	Electronic Security Devices
No. BP59	Second Book of CMOS IC Projects
No. BP65	Single IC Projects
No. BP69	Electronic Games
No. BP71	Electronic Household Projects
No. BP74	Electronic Music Projects
No. BP76	Power Supply Projects
No. BP80	Popular Electronic Circuits – Book 1

50 PROJECTS USING
I.C. CA3130

by

R. A. PENFOLD

BERNARDS (Publishers) LTD
The Grampians
Shepherds Bush Road
London W6 7NF
England

Although every care is taken with the preparation of this book, the publishers or author will not be responsible in any way for any errors that might occur.

©1977 BERNARDS (Publishers) LTD

I.S.B.N. 0 900162 65 1

First Published February 1977
Reprinted July 1980

Printed and Manufactured in Great Britain by
C. Nicholls & Co. Ltd.

CONTENTS

CHAPTER 1 — ABOUT THE I.C.   Page
   Operational Amplifiers 1
   Performance Figures 3
   Handling the Device 4
   Offset Null 5

CHAPTER 2 — AUDIO PROJECTS
   Inverting Amplifier 7
   Non-Inverting Amplifier 9
   A.F. Only Version 10
   Buffer Amplifier 12
   Bootstrapping 13
   Magnetic Cartridge Preamp 14
   Peak Level Indicator (P.L.I.) 15
   P.L.I. with Hysteresis 17
   Stereo P.L.I. 18
   Tone Control Circuit 21
   Audio Mixer 21
   500mW Audio Amplifier 24
   Higher Output Power Version 25
   Low Power Amplifier 25
   Compression Amplifier 25
   Hysteresis 29

CHAPTER 3 — R.F. PROJECTS
   M.W. Radio 31
   Crystal Calibrator 32
   R.F. Generator 34
   A.F./R.F. Signal Injector 35
   A.F./R.F. Signal Tracer 38

CHAPTER 4 — TEST EQUIPMENT
   High Impedance Voltmeter 41
   Resistance Meter 43
   Sinewave Generator 45
   A.G.C. Sinewave Generator 47
   Calibration 49
   Sine to Squarewave Converter 49
   Squarewave Generator 51
   Low Current P.S.U. 51
   High Current Version 54
   Continuity Tester 55

CHAPTER 5 — HOUSEHOLD PROJECTS
   Metronome 57
   Rain Alarm 58
   Light Switches 59

    Latching Circuits 62
    Sound Activated Switch 65
    Latching Version 67
    Christmas Tree Lights Flasher 69
    Simple Organ 69

CHAPTER 6 — MISCELLANEOUS PROJECTS
    Touch Switch 75
    Flashing Pilot Light 77
    Analogue Stopclock 78
    Over Voltage Protection 80
    Under Voltage Shutdown 82
    Voltage Indicators 82
    Morse Practise Oscillator 84
    Electronic Heads or Tails 87
    White Noise Generator 89
    Current Limiter 91
    High Current Version 93
    Over Current Indicator 93

COMPONENTS 95

Chapter 1
ABOUT THE I.C.

Operation amplifier integrated circuits have become increasingly popular in circuits for the amateur electronics enthusiast. The reason for this is not hard to discover, and is simply that these are probably the most versatile type of semiconductor device currently available. They are also among the least expensive of integrated circuits, and often have an economic advantage over alternative circuit elements.

The CA3130 is manufactured by R.C.A., and at the time of writing it is a relatively new device. It is not as inexpensive as certain other popular operational amplifiers, such as the 741C and 748C types, but it is a more advanced in design than its less expensive rivals. This means that it is often capable of a higher level of performance than other devices, and that fewer discrete components are needed. This tends to offset its cost disadvantage.

If past experience is anything to go by, the cost of the CA3130 is likely to decrease with passing time anyway.

Operational Amplifiers

Many people may not know exactly what an operational amplifier is, and so a brief description is provided here. Details of how an operational amplifier can be used in practical amplifiers, and other circuits, will not be given here, as this will be explained in the sections which come later. These provide detailed information of many circuits for diverse applications.

A theoretically perfect op. amp. has an infinite voltage gain, infinite input impedance, zero output impedance, infinite bandwidth, and is capable of giving a peak to peak output voltage swing which is equal to the supply rail potential. The circuit has two inputs, and these are termed the inverting input, and the non-inverting input. The circuit symbol for an op. amp. is shown in Figure 1.

If the non-inverting input is made positive of the inverting one, the output of the amplifier will swing positive. If the non-inverting input is negative with respect to the inverting one, the output swings negative. In a theoretically idealised op. amp. any difference in potentials between the two inputs will be enough to send the output fully positive or fully negative, but of course, no practical amplifier can achieve theoretical perfection for this parameter. Neither can it achieve theoretical perfection in any of the other parameters listed earlier, but most modern devices come close enough to be regarded as perfect in most respects. For instance, most op. amps. have a voltage gain of something like 100,000 times, and the typical figure for the CA3130 is some 900,000 times.

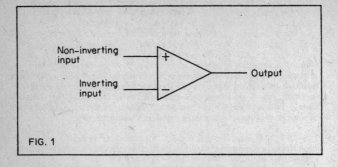

FIG. 1

In a few switching applications this full gain is required, but in all circuits needing linear amplification, this gain is greatly reduced by the application of negative feedback.

One parameter in which many well known devices fall short of theoretical perfection is that of input impedance. Bipolar transistors have relatively low input impedances, and since these form the basis of the input circuitry of most op. amps., this shortcoming exists. An example is the 741C I.C. which has a typical input impedance of 2Meg. with a minimum figure of 300k. This is not high enough for many applications, and even though the input impedance is increased to quite a large extent by the utilisation of negative feedback, the input impedance may still be too low.

Two chip op. amps have been available for some time, and these use a f.e.t. input stage on one chip, and the remaining circuitry is contained on a second chip. F.E.T.s have extremely high input impedances, and these two chip devices achieve input impedances of thousands of Meg. ohms. However, this is achieved at a price which puts them beyond the use of most amateurs, the actual cost being something like ten times that of a 741C I.C.

The CA3130 is manufactured using new techniques which enable the f.e.t. input stage and the main bipolar circuitry to be contained on a single chip. It is far less expensive than the two chip I.C.s, and is a very practical proposition. It uses a CMOS (complementary metal oxide semiconductor) input stage which has a voltage gain of only about five times. This is followed by two bipolar amplifying stages, the first having a voltage gain of 6000 and providing most of the units gain. The second is a Class A output stage which has a voltage gain of about 30 times.

Some operational amplifier I.C.s have internal compensation components, but the CA3130 does not. The purpose of the compensation circuitry is to reduce the upper frequency response of the device and so prevent it from becoming unstable. When used at low gains quite a high degree of high frequency roll-off must be used, but when used at high gains little or no roll-off is needed. Thus, if internal compensation is used, this must provide enough high frequency attenuation to prevent instability at low gains. This limits the bandwidth of the device unnecessarily when it is used at comparatively high gains.

Therefore using external compensation is not really a disadvantage even if it does slightly increase the number of discrete components required. It enables the bandwidth of the device to be optimised for any level of voltage gain. In the case of the CA3130 only a single loss value capacitor is used to provide the necessary frequency compensation.

Performance Figures

As will be seen from the main performance figures of the CA3130, which are given below, this device has a high level of performance.

Input impedance	1.5 million Meg. ohms.
Open Loop Voltage Gain (the gain without negative feedback)	900,000 times
Input Bias Current	5pA (1pA = 1 millionth of a micro Amp.)
Gain-Bandwidth Product	15MHZ
Slew Rate	30V/micro sec.
Operating Temperature Range	−55 to +125 degrees C
Supply Voltage Range	5V. to 16V., or a balanced positive and negative supply of ±2.7V. to ±8V.
Current consumption from 9V supply with output at half supply voltage	2.5mA

The above are all typical ratings.

The CA3130 is contained in a TO-5 8 lead metal encapsulation, and its leadout diagram is shown in Figure 2. There are several versions of the I.C., and the CA3130T and CA3130S versions are the ones that are required for the circuits described in this book. The CA3130T has a standard TO-5 can and leadouts whereas the CA3130S has its leadouts formed into an 8 pin dual in line configuration. These two devices are electrically identical.

Other versions of the CA3130 have a more rigid specification in some respect or other, and these will work in these circuits. They are however, more highly priced than the two basic versions.

One advantage over this device when compared to most other op. amps. is that when lightly loaded, the output can swing to within a matter of a few millivolts of either supply line. Most other devices can only manage an output swing (peak to peak) of about 4 volts less than the supply voltage. This enables the CA3130 to be used in simple circuit

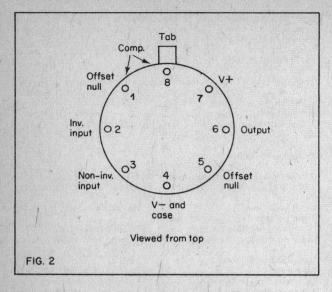

FIG. 2

configurations which would not be possible using most other op. amps.

Handling The Device

As many readers will be aware, CMOS devices can be damaged quite easily if they are subjected to high voltage static charges, and precautions must be taken not to destroy them due to careless usage and handling. The CA3130 is not as easily damaged as some CMOS devices. This is partly due to the fact that only the input circuitry is of the CMOS variety, but also there are zener protection diodes incorporated in the I.C.

Even so it is adviseable to take reasonable care when using and handling these I.C.s. Usually the devices are supplied with their leadout wires pushed into a piece of conductive foam. They should be left in this until it is time to connect the device into the rest of the circuit. The I.C. should be the last component to be soldered into circuit.

Use a soldering iron with an earthed tip when connecting the device. It is not a good idea to apply an input signal to any semiconductor device when the power supply is not connected, and the CA3130 is no exception to this.

Offset Null

Most op. amps. have an offset null facility, and the CA3130 is one of these devices. The purpose of the two offset null leadouts is to enable the output to be adjusted to zero (in the case of a dual supply), or to half the supply voltage (in the case of a single supply) even though the input terminals are not at quite the same voltage.

This is a useful feature, but it is not required in any of the circuits described here.

Chapter 2
AUDIO PROJECTS

Perhaps the most obvious use for an op. amp. is in audio amplifier circuits, and they perform this function very well indeed, even though they were not designed to perform this task. Op. amps. are actually intended for use in analogue computers where they carry out mathematical operations. Hence the name 'operational amplifier'.

It will be helpful to first consider how an op. amp. would normally be connected, before considering its use in a number of audio amplifying applications. An op. amp. is really intended to be used as a D.C. amplifier, and it is this fact that is largely responsible for its extreme versatility. There are two basic amplifying modes in which an op. amp. can be employed, the inverting amplifier and the non-inverting amplifier. These two basic configurations are shown in Figure 3 and Figure 4 respectively.

Note that these circuits are not run from the usual single supply, but need dual supplies of equal voltage. One is negative of the central earth point, and the other is positive of earth. Thus both the inputs and the output can have either polarity with respect to earth.

Inverting Amplifier

The inverting amplifier is the most simple of the two arrangements, and this requires only two external resistors plus the compensation capacitor (C1). It is the values of the two discrete resistors which determines the gain and input impedance of the amplifier.

Designing a circuit for a given input impedance and voltage gain is very simple, and even a complete beginner should have no trouble with this. The input impedance is equal to the value given in R1, and this is multiplied by the required voltage gain to find the required value for R2. Thus if an amplifier with an input impedance of 10k and a voltage gain of 100 times (40dB.) is required, R1 would have a value of 10k, and R2 would have a value of 1Meg. (10,000 x 100 = 1000,000 ohms, or 1Meg.). It is this simple relationship between resistor values and voltage gain that enables the circuit to be used for mathematical operations.

What is happening in this circuit is really quite simple to understand. Assume that R1 has a value of 10k and R2 has a value of 1Meg. If an input signal of 1mV. is applied to the circuit (and it is positive with respect to earth), this will take the inverting input positive of the non-inverting one. This will cause the output to swing negative as the inputs have become unbalanced. The output will in fact swing 100mV. negative. R2 has a value which is 100 times higher than that of R1, but it then

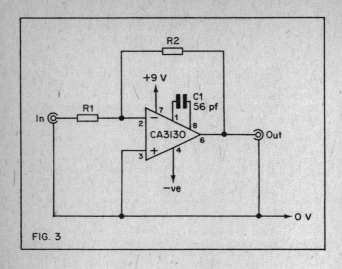

FIG. 3

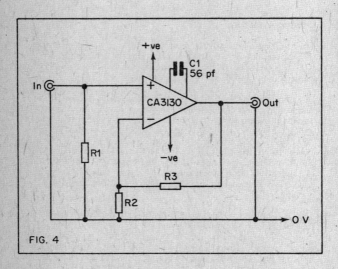

FIG. 4

has 100 times the voltage developed across it. Therefore the equal positive and negative currents through R1 and R2 respectively cancel one another out, and the inverting input will be at the same potential as non-inverting input. It is important to remember that the op. amp. has an extremely high gain, and that only a minute difference is needed between the two inputs in order to send the output fully positive or negative. Virtually all circuits using operational amplifiers rely upon this fact for their correct operation.

We have seen above that this circuit will always try to maintain the two inputs at the same voltage, and how this enables the discrete resistors to set the required voltage gain. It is also this that makes the input impedance of the circuit equal to the value of R1. It will be seend from Figure 3 that the non-inverting input of the I.C. is connected to earth. Therefore the circuit is stabilising the inverting input at earth poential, and is creating what is termed a 'virtual earth'. Thus the impedance between the input and the virtual earth is equal to the impedance of R1, and so the value of R1 sets the input impedance of the circuit.

Non-Inverting Amplifier

As its name implies, an inverting amplifier has input and output signals of opposite polarity, as we have just seen. This is not always convenient, and it is sometimes necessary to use the non-inverting amplifier configuration. This is again very simple with three discrete resistors being used to set the required voltage gain and input impedance.

Here R1 sets the required input impedance, and the voltage gain of the circuit is determined by R2 and R3. The input impedance is equal to the value of R1, and this is obviously so since this resistor is the only path between the input and earth. Remember that the input impedances at the inputs to the I.C. are theoretically infinite (and in the case of the CA3130 they are so high that for all practical purposes they can be considered as infinite).

R2 and R3 set the voltage gain of this circuit in much the same way as they did in the previous one. The non-inverting input is not directly connected to the earth line, and so the voltage gain mathematics are not quite the same as in the previous circuit.

Again the circuit will try to maintain the same potentials at the two inputs. If R2 has a value of 10k and R3 has a value of 1Meg., and we again assume there is a positive input of 1mV., the output will swing positive until the voltage on the inverting input is 1mV. positive. R3 and R2 act as a simple potential divider and 101mV. will be needed at the output to produce 1mV. at the inverting input. Thus the circuit has a voltage gain of 101, and not of 100 as was the case in the previous circuit.

The equiation that gives the voltage gain of this circuit is R2 + R3 divided by R2. Incidentally, R1 and R2 of Figure 3, and R3 and R2 of

Figure 4 form what are usually termed a feedback loop, or sometimes it is alternatively called a feedback network.

One point which anyone thinking of designing a simple amplifier of either type should bear in mind is that although the resistor values can be as high or as low as you like in theory, in practise this is not the case. Theoretically the op. amp. has an output impedance of zero, but in a practical circuit this is usually something in the order of a few hundred ohms. Therefore, if the feedback resistors are very low in value the output impedance of the device will upset the operation of the circuit.

Also there is an upper practical limit to the values which can be employed. For one thing very high value resistors are simply not available. Also, the higher these values are made, the greater the susceptibility of the circuit to stray pick-up of mains hum and other sources of electrical interference.

In the case of the non-inverting amplifier there is also the problem of feedback between the output and input of the circuit. This is coupled through stray circuit capacities that inevitably exist in any circuit. Any such feedback is usually very low unless a very careless component layout is used. However, if a non-inverting amplifier is used with a combination of high gain and high input impedance, even an extremely modest amount of stray feedback will be sufficient to cause the circuit to break into oscillation.

If such a combination of gain and input impedance is required, it is better to use a separate high input impedance buffer amplifier feeding the main (low input impedance) voltage amplifier. This enables the input and output to be physically well separated, and this reduces the stray feedback to an insignificant level.

The above also applies to the following two variations on the basic amplifiers.

A.F. Only Versions

Although primarily intended as D.C. amplifiers, the configurations of Figures 3 and 4 can also be used as A.F. amplifiers. The use of a dual supply is rather inconvenient though, and by using a few additional discrete components it is possible to devise a circuit that only needs the normal single supply.

These circuits are only suitable for A.C. applications, and they are shown in Figures 5 and 6. They are inverting and non-inverting amplifiers respectively.

If we consider Figure 5 first, the similarity between this and Figures 3 should be quite apparent. There are only two real differences. Instead of using a dual supply, the potential divider formed by R3 and R4 is used to provide a sort of centre tap on the supply lines. C4 is a bypass

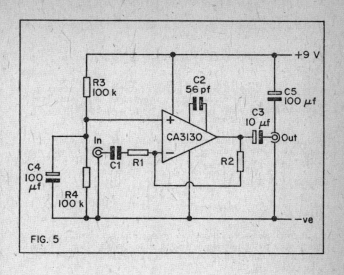

FIG. 5

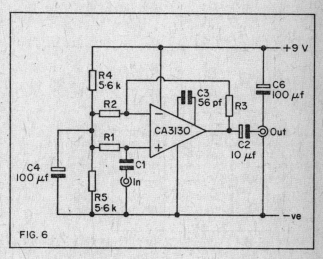

FIG. 6

capacitor. D.C. blocking capacitors are needed at the input and output, and these are C1 and C3 respectively.

The same modifications have been carried out on Figure 4 to produce the circuit of Figure 6.

The gains and input impedances of these circuits are determined in exactly the same way as the previous two designs. The value of the input coupling capacitor (this is C1 in both Figure 5 and Figure 6) must be varied to suit the input impedance of the circuit.

With an input impedance of 1k a value of 10mfd. is suitable for C1. For higher input impedances the value of C1 should be reduced proportionately. For instance, it would be 2mfd. for an input impedance of 5k, 0.1mfd. for an input impedance of 100k, and 5nf for an input impedance of 2Meg.

Buffer Amplifier

Apart from use in general purpose preamplification circuits, op. amps. can be used very effectively in several more specialised applications. The buffer amplifier shown in the circuit diagram of Figure 7 is a good example of this.

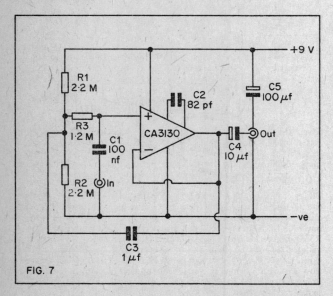

FIG. 7

The idea of this type of circuit is to provide an input impedance which is as high as possible. This type of circuit is often employed in items of test equipment, such as A.C. millivoltmeters, where it is desirable for the test gear to draw an absolute minimum of power from the circuit under test. The circuit has a voltage gain of unity, its sole purpose being to provide an ultra high input impedance, and to provide any voltage gain that may be required. The input and output signals are in phase.

To obtain unity gain from an operational amplifier is very simple indeed. It is just a matter of connecting the inverting input to the output. The voltage at the output will then be identical to the input voltage at the non-inverting input, provided the input signal does not go outside the output voltage swing limits of the op. amp. When connected in this way the circuit has what is termed a 100% negative feedback loop between its inverting input and its output.

In this circuit the I.C. has such a feedback loop. R1 and R2 provide a bias voltage of half the supply potential at their common connecting point, and this voltage is fed to the non-inverting input of the I.C. via R3. Thus under quiescent conditions the input and output of the device are at about half the supply voltage.

The input signal is coupled to the I.C. by way of D.C. blocking capacitor C1, and the low impedance output signal is obtained via D.C. blocking capacitor C4.

Bootstrapping

C3 is a bootstrapping capacitor, and it is this that gives the circuit its extremely high input impedance. If C3 is left out of circuit, R1, R2, and R3 will shunt the very high input impedance of the I.C. and will produce an input impedance to the amplifier as a whole of about 2.3Meg. This is adequate for many purposes, such as a preamplifier for a crystal or ceramic cartridge, and in such cases C3 can be omitted.

Theoretically, including C3 boosts the input impedance of the amplifier to the same input impedance as the I.C. possesses. In practice there are losses in the circuit which result in less than 100% effectiveness, but the input impedance of the test circuit was far too high for the author to be able to measure it, and was at least some hundreds of Meg. ohms.

The operating principle of the bootstrapping technique is quite simple to understand. Assume that there is a positive going input signal which raises the input potential by one volt, the output will also increase by one volt. This increase is coupled through C3 to the junction of R1 to R3. Therefore the rise in potential at the junction of C1 and R3 is matched by a similar rise at the opposite end of R3, and the current flow through the resistor will remain constant. None of the input signal can flow through R3, which in consequence appears as an infinite impedance to input signals. R1 to R3 are in effect non-existant as far as input signals are concerned.

Magnetic Cartridge Preamp.

A magnetic cartridge preamplifier must have very precise characteristics. It must provide an input impedance of 47k (occasionally a different figure is required), high voltage gain, and a well defined level of bass boost and treble cut. As the output from a magnetic cartridge is only a few mV., it must also have a low noise level if satisfactory results are to be obtained.

All these requirements are easily met by using a high performance op. amp. such as the CA3130. The circuit diagram of such an amplifier is shown in Figure 8.

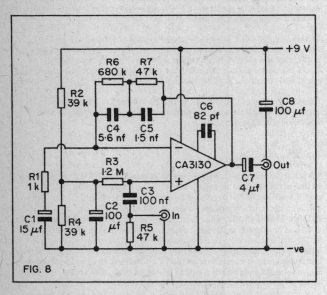

FIG. 8

This is basically a non-inverting amplifier with an input impedance of 1.2Meg. R5 is shunted across the input in order to reduce the input impedance to the required level. If a different impedance is required, the value of R5 can be adjusted accordingly. It may at first appear to be better simply to leave out R5 and reduce R3 to a value of 47k. This would work, but it would then be necessary to use a higher value for C3 in order to maintain a good bass response. This is undesirable since when the power is first applied to the circuit, C3 has to attain its normal working charge. This causes a pulse of current to flow through the coil of the cartridge, and over a period of time this can have detrimental effect on the cartridge.

The input circuitry used here enables a low value input coupling capacitor to be employed while still providing an excellent bass response. Some of the charge current for C3 passes through R5, and not through the cartridge, and there is no real risk of the cartridge being damaged when using this circuit.

Capacitors C4 and C5 are included in the feedback loop of the circuit, and these have an impedance which decreases with increasing frequency. Thus there is more feedback at treble frequencies than there is at middle ones. This provides the required treble cut. At bass frequencies the feedback capacitors have a very high impedance, and this results in a low level of feedback. This gives the necessary bass boost.

It is perhaps worth pointing out that a capacitor is shown connected across the supply lines in all the circuits featured in this book. This is a supply decoupling component and it should not be left out. It helps to prevent the circuit from becoming unstable. For maximum effect it should be mounted physically close to the I.C.s supply leadout wires, but in practice the author has not found this to be at all essential.

At a frequency of 1kHZ the circuit has a gain of a little under 100 times, and from most magnetic cartridges it will produce an output of around 300mV. This should be enough to drive most power amplifiers. If necessary, the gain can be increased slightly by reducing the value of R1 to some extent. On the other hand, if the gain is found to be excessive, a higher level of performance can be attained by slightly increasing the value of R1.

It is probable that most constructors of the unit will require stereo operation, and it is then merely necessary to build two of these circuits, one to handle each channel.

Peak Level Indicator

So far we have only considered the use of an operational amplifier in a linear mode. It can also be used very effectively as an electronic switch, even in some audio applications. The circuit diagram of the peak level indicator is shown in Figure 9, and in this the CA 3130 is used as a type of electronic switch. It is used as a comparator.

The purpose of a peak level indicator is to show when a signal being processed by a tape recorder (or some other piece of audio gear) has exceeded some predetermined level. The usual method of indicating the overload, and the method adopted here, is to have an indicator light briefly come on.

This type of circuit has an advantage over the usual VU type of level meter, although this unit is intended to supplement rather than replace a VU meter. The main drawback of a VU meter is the comparative slowness with which it responds to changes in input signal level. It

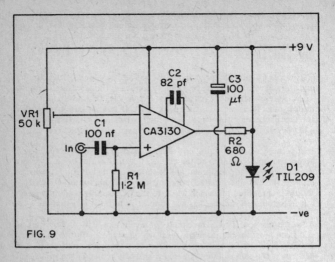

FIG. 9

will simply miss any sudden crescendos which could be overloading the equipment.

A VU meter is usually calibrated using a sinewave input signal. This has a peak level of about 1.4 times the R.M.S. level, whereas most passages of speech and music have a peak level of two to three times the R.M.S. level. Even on signals of a fairly constant level an average reading VU meter can produce misleading readings.

A peak level indicator can thus be a very useful adjunct to a VU meter as it responds to peak rather than average or R.M.S. levels, and it is virtually instantaneous in operation.

With reference to Figure 9, assume that the circuit is required to indicate any peak over a level of 3 volts. VR1 would then be adjusted so that its slider was one third of the way up its track, and there would be a potential of three volts at the inverting input of the I.C.

The input signal is applied to the non-inverting input of the I.C. by way of C1. Under quiescent conditions this input is biased to earth potential by R1, and so the input signal swings symmetrically either side of earth. Provided the peak level of the input signal does not exceed a level of three volts, the non-inverting input will obviously be at a lower potential than the inverting one, and the output will be fully negative. The light emitting diode indicator, D1, will therefore remain off.

If the input signal should ever exceed three volts in peak amplitude, then the output will swing fully positive for the duration that this threshold level is broken. Current will then be supplied to the L.E.D. indicator from the output of the I.C. through R2, and a visual indication of the overload will be produced. R2 is merely a current limiting resistor and it is required to protect the L.E.D. against passing an excessive current.

The unit has a high input impedance and it can be fed from any convenient point on the main equipment without adding significantly to the loading on the relevant stage. VR1 can be adjusted to produce an indication from any peak level from a few mV. to several volts.

Setting the unit up is quite simple, but a sinewave generator (or some other source of a sinewave signal) is really required. With the sinewave signal coupled to the input of the tape deck (or whatever), adjust the input level controls to produce a reading on the VU meter which is equal to the level at which the peak level indicator is to operate.

Start with VR1 adjusted for maximum voltage at its slider, and there should then be no indication from D1. Gradually adjust VR1 for a lower voltage at its slider, and eventually a faint but visible glow will be produced from D1. When this point is reached, VR1 has the correct setting.

P.L.I. With Hysteresis

Most peak level indicators work on basically the same principle as the circuit described above, but a few are a little more complex, and incorporate hysteresis. Many people may be unfamiliar with the term 'hysteresis', and all this means is that the circuit has a fast attack and slow decay. In other words, if the threshold level of the unit is exceeded the lamp will come on virtually instantaneously, but when the signal falls below the threshold level again, the lamp will remain on for a short time.

The reason for doing this is quite simple, and is that by doing so the unit will give a more clear indication of an overload of very brief duration. Some people therefore have a preference for this type of circuit. The circuit diagram of a peak level indicator with hysteresis appears in Figure 10.

The input circuitry is identical to that employed in the circuit of Figure 9, but the output is not used to directly drive the L.E.D. Instead it is coupled via C3 to a voltage doubling rectifier and smoothing network using D1, D2, and C4. The I.C. has a low output impedance and so C4 is very quickly charged to the peak level of the output signal. This turns on Tr1, which is used as an emitter follower. This has the L.E.D. indicator (D3) and the series current limiting resistor (R2) as its emitter load, and so the L.E.D. is illuminated when Tr1 turns on.

When the signal drops below the threshold level and there is no longer an output from the I.C., D3 will not turn off immediately. It must remain on until C4 has almost fully discharged into the base of Tr1, and this takes about 0.25 seconds. The turn off of D3 is thus delayed slightly, with even very brief overloads producing an output of 0.25 seconds duration.

This circuit is connected and set up in exactly the same way as the previous one.

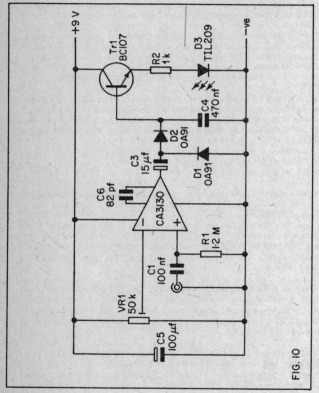

FIG. 10

Stereo P.L.I.

For stereo operation it would of course be possible to use two peak level indicators, one to monitor each channel. Most commercial equipment seems to use a single indicator lamp to indicate an overload on either or both channels, however. There is quite a good case for doing this, as with two indicator lamps and two meters to monitor, using separate lamps for each channel could make the equipment rather demanding to use.

It is possible to convert two of the peak level indicators just described (either type) to feed a single indicator lamp by using the simple circuit shown in Figure 11. The two peak level indicator circuits have their L.E.D.s and series current limiting resistors omitted. The outputs of the I.C.s are fed instead to the inputs of this circuit.

The device operates in much the same way as the P.L.I.s themselves. The inverting input of the I.C. is connected to the potential divider formed by R1 and R2, and so roughly one quarter of the supply potential is present at this input. In order to obtain a positive output from the I.C, and so get D1 to light up, the non-inverting input must be taken above the inverting one in potential.

If the outputs from both P.L.I. circuits are low, this will not be achieved and the L.E.D. will be in the off state. If either P.L.I. circuit has a output in the high state, R3 and R4 will form a potential divider with one input at virtually the negative supply voltage, and the other at almost the same voltage as the positive supply. This causes about half the supply potential to appear at the non-inverting input of the I.C, and so D1 is switched on. It does not matter which output is high and which is low, the effect is the same in either case.

If both inputs are high, the non-inverting input will be at virtually the positive supply rail potential, and again D1 will be illuminated. This

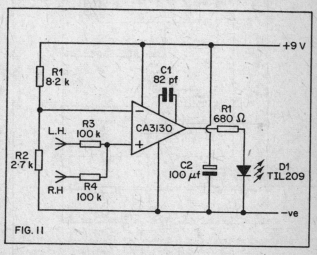

FIG. 11

circuit thus provides the necessary action, as if either or both the P.L.I. outputs are high, D1 comes on.

The I.C. is in fact being used as a simple logic circuit. It is, in effect, a positive two input OR gate, as the output is positive if either the L.H. OR R.H. inputs are positive.

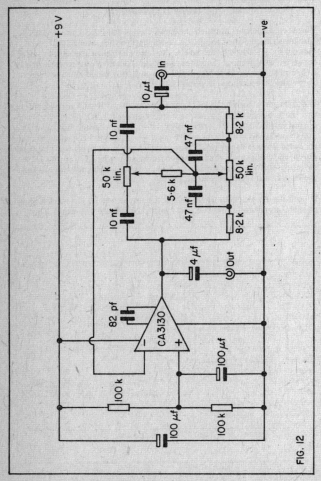

FIG. 12

Tone Control Circuit

We have already seen in the case of the magnetic cartridge preamplifier how it is possible to alter the frequency response of an amplifier by the use of frequency selective negative feedback. The same method can be used to produce a high quality active tone control.

An active tone control has the advantage over a simple passive circuit that it does not need to have large losses between the input and output. When adjusted for a flat frequency response a passive tone control circuit usually has an insertion loss of around 20dB. (10 times).

The circuit diagram of the active tone control is given in Figure 12, and when adjusted for a flat frequency response this circuit has a voltage gain of approximately unity. It offers the usual bass and treble boost and cut, and it has a very low level of noise and distortion.

Audio Mixer

An op. amp. makes the ideal basis for a multichannel audio mixer, and the circuit of such a unit is shown in Figure 13. This is basically an inverting amplifier, but as will be seen from the circuit diagram, each input channel has its own input resistor.

If we only consider input 1 for the moment, then the circuit operates as a straight forward inverting amplifier with R1 and R6 setting the voltage gain at unity. C1 is the D.C. blocking capacitor at the input and VR1 is the fader control for channel 1.

When two or more inputs are being used the circuit acts as a sort of current adding circuit. For the sake of this explanation we will assume that there is an input voltage of 10mV. at each of the three inputs. In order for the output to balance the combined currents through R1, R2, and R3, and maintain the I.C.s inputs at the same levels, it must swing 30mV. negative. It will be apparent from this that the voltage at the output is the sum of the input signals, and in this way the circuit operates as a high quality mixer.

It is a simple matter to construct a stereo version of this circuit, and all that is required is two separate circuits, one for each channel. The circuit is shown with three inputs, but it will work with as many inputs as one wishes, simply by adding extra fader controls with the appropriate input resistors and D.C. blocking capacitors. The unit is intended to be connected in a fairly high level part of the overall set up, with any pre-amplification that is needed being added ahead of each input. However, if required, the gain of the unit can be increased by increasing the value of R6. The voltage gain is equal to the value given to R6 divided by 100k.

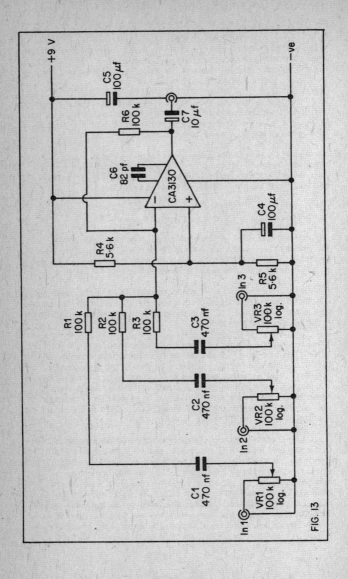

FIG. 13

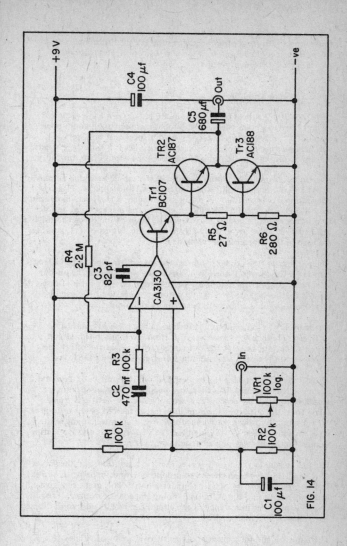

FIG. 14

In common with most other op. amps., the CA3130 is capable of providing extremely low levels of noise and distortion, and this circuit provides really excellent results.

500mW. Audio Amplifier

The CA3130, like most other op. amp. I.C.s, does not have sufficient output drive to obtain adequate volume if it is used as an audio power amplifier. The device can be used to drive a speaker if only very low volume levels are required, but for most purposes it is necessary to add a discrete output stage in order to increase the output drive capability.

The circuit diagram of a 500mW. audio power amplifier which employs a CA3130 in the preamplifier section is shown in Figure 14. TR1 is a conventional emitter follower driver stage having R6 as its main emitter load. R5 is connected in this emitter circuit, and it is required in order to produce a small standing bias across the bases of the output transistors, Tr2 and Tr3. These are directly driven from the output of Tr1, and they are used in the emitter follower mode also.

The purpose of R5 is to help reduce cross-over distortion to an insignificant level. Cross-over distortion occurs at the point where the output signal changes polarity, and it is a particularly annoying form of distortion since it is at its highest level at low output powers. Distortion tends to be more noticeable at low volumes than at high ones.

Normally R5 would need to be much higher in value in order to sufficiently reduce the cross-over distortion level, but in this circuit a large bias current through the output transistors is unnecessary. This results from the use of large amounts of overall negative feedback.

The voltage gain between the base of Tr1 and the emitter of Trs2 and 3 is almost exactly unity. It is therefore possible to use an op. amp. preamplifier to drive the base of Tr1, with the negative feedback being taken from the output of the amplifier, rather than from the output of the op. amp. By using negative feedback over the entire circuit in this way, the unit achieves a very high level of performance, and is superior to most simple battery powered audio amplifiers.

The I.C. is used in the inverting amplifier configuration, and the values of R3 and R4 have been chosen to produce an input impedance of 50k and an input sensitivity of a little under 100mV. This input sensitivity is for an output of 500mW. into an 8 ohm impedance speaker. If required, the gain and input impedance of the circuit can be altered by modifying the values of VR1, R3, and R4.

The required gain for a given input sensitivity can be calculated by dividing 2,000 by the required input sensitivity in milli-volts. R3 and VR1 should have about the same value, and the input impedance of the unit is then approximately equal to R3 divided by two.

It is not a good idea to use a speaker of less than 8 ohms impedance as this will result in the circuit being rather inefficient with regard to current consumption, and will also result in some loss of output quality. It is quite safe to do so however, and no damage to the circuit is likely to result from this. Higher impedance speakers can be used, but there will be a lower maximum output power available if this is done. For example, an output power of only about 100mW. will be possible if an 80 ohm speaker is used.

Higher Output Power Version

By using a higher voltage power supply it is possible to obtain a substantially higher output from this circuit. It will give almost 2 watts from a 12 volt supply and about 3 watts from a 15 volt supply. A supply of significantly more than 15 volts should not be used. The only circuit value that needs any adjustment is R5. This must be reduced to 18 ohms for a 12 volt supply, and 12 ohms for a 15 volt supply.

It will also be necessary to fit the output transistors with clip on heatsinks, and the heatsinks should be bolted to an area of about 100 sq. c.m.s of 16 s.w.g. aluminium. This ara should be regarded as a minimum, and more substantial heatsinking should be provided if space permits this. Inadequate heatsinking will almost certainly result in the destruction of the output transistors due to overheating.

The higher power versions of the circuit should be used with 8 ohm speakers, and definitely not with speakers having an impedance of less than 8 ohms.

Low Power Amplifier

In applications where only a very low output power is satisfactory, the CA3130 can be used to drive a speaker without using a discrete output stage. The circuit of Figure 15, for instance, has an output power of only a few milli-watts, but it will produce sufficient volume for use in the audio stages of a bedside receiver, of any similar application.

This is really just an inverting amplifier driving a high impedance speaker.

Compression Amplifier

An audio compression amplifier is a circuit that does not have a fixed voltage gain, but instead has a gain that varies with level of the input signal. On low level signals the amplifier develops its full voltage gain, but when a certain input level is exceeded the gain of the amplifier begins to decrease. The higher the input is taken above the threshold level, the lower the gain of the amplifier becomes.

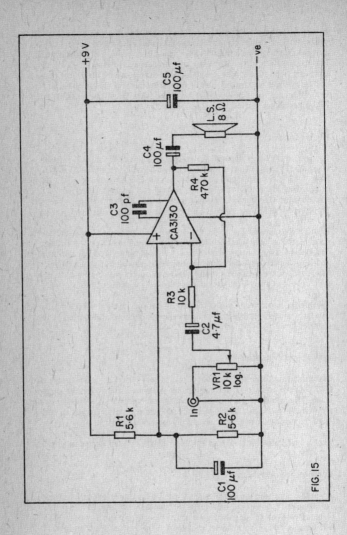

FIG. 15

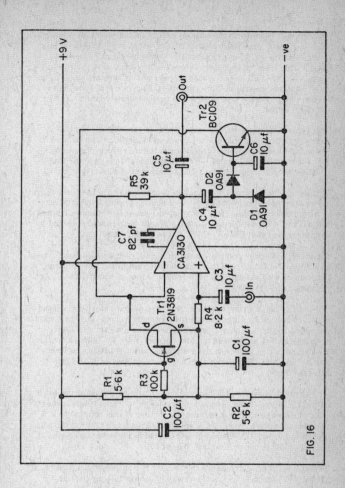

FIG. 16

The result of this is that the output level of the amplifier does not go above a certain amplitude, and signals above the input threshold level generate an output level at or slightly below this output amplitude. Thus a wide range of input signal levels will produce a virtually constant output level. For this reason circuits of this type are sometimes referred to as constant volume amplifiers. They are also called compression amplifiers as they reduce or compress the dynamic range of the input signal.

This type of circuit is used in many fields of electronics. In radio communications for instance, a compression amplifier can be used in the audio stages of a transmitter. Here it has the effect of ensuring that the modulation level of the transmitter is always close to 100%, and that maximum use of the transmitters output power is achieved.

Circuits of this type are often used in tape recording in the form of an automatic recording level facility, or a peak recording level limiter. A compression amplifier is a sort of audio automatic volumes control circuit, and it is possible to use such a circuit in the audio stages of a simple receiver that has no other form of A.V.C. circuit.

Figure 16 shows the complete circuit diagram of the audio compression amplifier. The I.C. is used as a non-inverting amplifier, and its gain is determined by the feedback network consisting of the drain to source resistance of Tr1, and R5.

Tr1 is a field effect transistor, and this type of transistor acts as a simple resistor between its drain and source terminals when a fairly low voltage is applied to them. Under quiescent conditions the gate and source terminals of Tr1 are held at the same potential by R3. When a f.e.t. is connected in this fashion it has a drain to source resistance of only about 100 ohms. The circuit thus achieves a voltage gain of a few hundred times.

Any input signal fed to C3 thus appears greatly amplified at the output of the I.C. Some of the output signal is taken to the output socket via D.C. blocking capacitor, C5, but a proportion of it is coupled to a rectifier network by way of C4. D1 and D2 form the rectifier circuit, and their output is developed across C6 which smooths the rough D.C. output of the rectifiers to a D.C. bias voltage.

This voltage is proportional to the level of the output signal. If it is about 0.6V. or less it will not have any effect on the circuit, but if it exceeds this level it will begin to turn on Tr2. As Tr2 begins to conduct, it takes the gate potential of Tr1 down towards the negative supply rail, and this gives Tr1 a reverse bias. This greatly increases its drain to source resistance, and in consequence it reduces the gain of the amplifier. The higher the amplitude of the output signal, the more the gain of the circuit is reduced.

In this way the output tends to be stabilised at a certain level. Measurements made on the prototype show that this circuit has a very effective compression effect. On input signals up to about 1mV. the gain of the amplifier is constant at about 260 times. Thus an input of 1mV. R.M.S. produces an output of 260mV. R.M.S. As the input signal level is increased about 1mV., the gain of the circuit rapidly diminishes. With an input of 70mV. an output level of approximately 310mV. is obtained. Therefore increasing the input signal above the 1mV. threshold level by a factor of 70 only increases the output level by about 20%!

At an input of around 70mV. the compression effect becomes saturated with Tr2 being turned on as far as it can be. Increasing the output above this level therefore results in a rapid rise in the output level, and also the output signal will be found to be very distorted. The input signal level must therefore be limited to 70mV. or less, and if necessary an attenuator must be added at the input to accomplish this. Provided the input signal is kept below the saturation level, an extremely low distortion level is obtained.

Hysteresis

Like the second of the two peak level indicator circuits that were described earlier, this circuit has hysteresis. This is almost essential, as the circuit would obviously be of little use if it only responded very slowly to increases in input level. The slow decay time of the unit is also an important feature, as this prevents the gain from rising during brief pauses between words, or during brief lulls in music.

The circuit has a decay time of about 1 to 2 seconds, but this can be varied if necessary by altering the value of C6. Reducing its value reduces the decay time proportionally, and increasing its value lengthens the decay time accordingly.

Care must be taken when using this circuit as its has quite a high voltage gain, and the input and output are in phase. A careless component layout could easily lead to instability. For the same reason, and to avoid stray pick up of mains hum etc., all leads at the input should be screened.

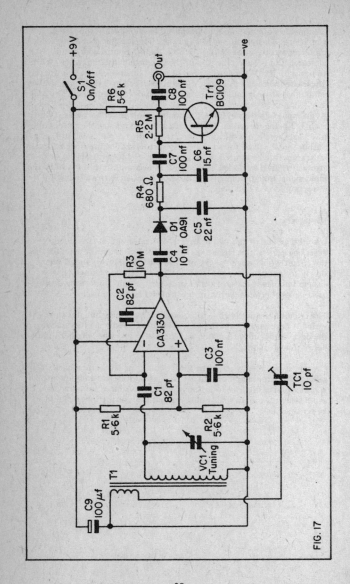

FIG. 17

Chapter 3
R.F. PROJECTS

Operational amplifiers are not generally associated with R.F. applications, and apart from a few special types, they are not really intended for this type of use. However, the CA3130 has quite a wide bandwidth and it has been found to be possible to use it in a few R.F. circuits.

M.W. Radio

This is rather a novel application for an operational amplifier, and is one that the author has not come across before. The circuit diagram of the 'M.W. Radio' appears in Figure 17.

The I.C. is used as a sort of inverting amplifier, with the signals from the ferrite aerial being fed to the inverting input of the I.C. via D.C. blocking capacitor C1. In order to obtain a reasonable level of gain and selectivity from the circuit it is necessary to use regeneration. This is applied from the output of the I.C. to the smaller (untuned) winding on the ferrite aerial by way of TC1. The level of regeneration is adjusted using TC1.

VC1 is an ordinary tuning control.

Although regeneration is used over the R.F. stage of the receiver, this is not used as a regenerative detector, and the main output of the I.C. is coupled via C4 to the diode detector, D1. C5, R4, and C6 are an R.F. filter. The audio output from this filter is fed to a single transistor audio amplifier stage using Tr1. This is a common emitter amplifier, and it provides a low noise level and a high voltage gain.

The output can be used to drive a crystal earphone, high impedance magnetic headphones, or any other medium to high impedance load.

Most of the circuits in this book can be constructed quite satisfactorily using Veroboard as a very convenient constructional basis. It would be advisable to use a different technique for this circuit, and for the two which follow, as stray capacitances must be kept to a minimum. They could otherwise be of sufficient level to upset the operation of these circuits causing either instability of loss of performance. P.C.B. or plain matrix boards are suitable constructional forms for these three circuits.

T1 can be any ferrite rod for use in transistorised circuits, and the value of VC1 must be chosen to suit the particular ferrite rod aerial used. Suitable components for use in these positions are a Denco MW/5FR for T1 and a Jackson type 01 208pf variable capacitor for VC1.

For optium results to be obtained from the set it is necessary to adjust TC1 very precisely. With this initially set for about minimum capacitance it should be found that a few stations can be received, although they will probably be rather weak at present. Adjusting TC1 for increased capacitance will increase the regeneration level and should produce improved results. If no signals can be received, or this adjustment is found to have the opposite effect to that just described, it probably means that the phasing of T1 is incorrect. This can be rectified by reversing the connections into the feedback winding (i.e. the smaller of the two coils)

There is a limit to the amount of regeneration that can be effectively applied to the circuit, and if this limit is exceeded the set will break into oscillation at some point on the tuning dial (probably at the high frequency end). This will be hears as a tone of varying pitch as the set is tuned over a station, and proper reception will not be possible.

Adjust TC1 to insert as much capacitance into circuit as possible, without the set starting to oscillate at any setting of the tuning control. The unit will then have optium sensitivity and selectivity.

With most makes of ferrite aerial it is necessary to slide the coil assembly up and down the ferrite rod in order to find the position that gives the correct frequency coverage. The coil is then taped in that position.

Crystal Calibrator

A crystal calibrator is very useful for checking the calibration of a communications receiver, or for calibrating a home made S.W. receiver. Basically all it consists of is a crystal oscillator having an ouput which is rich in harmonics. The harmonics are the multiples of the fundamental frequency, and so a 1MHZ crystal calibrator will provide harmonics at 2MHz, 3MHz, 4MHz, etc. It is these harmonics that are used for calibration purposes, the fundamental frequency not being with the S.W. frequency spectrum.

The circuit diagram of the crystal calibrator is shown in Figure 18. This will work with virtually any crystal having a frequency between a few tens of kHz and several MHZ. It can thus be used with the usual 100kHZ and 1MHZ calibration crystals. The prototype circuit provided strong harmonics throughout the S.W. bands to beyond 30MHZ, and the circuit is therefore suitable for use with any S.W. receiver, regardless of frequency coverage or sophistication. Some crystal oscillators are reluctant to oscillate unless a high activity crystal is employed, or have a poor harmonic output unless such a crystal is used, but this circuit seems to give a strong output from any crystal that is in reasonable condition.

The op. amp. is used as a form of non-inverting amplifier, with the crystal and TC1 providing a positive feedback loop between the output and non-inverting input of the I.C. A crystal has two resonant frequencies, and these are called the series resonant frequency and the parallel resonant frequency. These are usually only a few hundred HZ

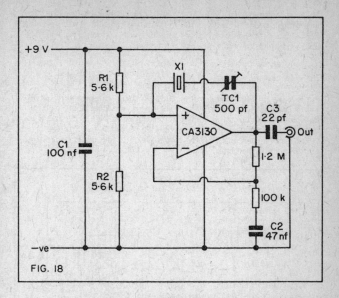

FIG. 18

apart, with the series resonant one being the lower of the two. At the parallel resonant frequency the crystal exhibits a very high impedance and at series resonance it has a very low impedance. At other frequencies a crystal has a fairly high impedance.

Thus in this circuit there will be a high level of feedback at series resonance, and very little feedback at other frequencies. The unit therefore oscillates at the series resonant frequency of X1.

A tight coupling will probably not be needed between the receiver and the calibrator as the output of the unit is quite strong. It will probably be sufficient to connect a lead to the output of the unit and then place this near the aerial socket of the receiver.

TC1 can be used to accurately set the unit up against a standard frequency transmission, such as the one to be found at 5MHZ. With the receiver adjusted to receive this transmission the output of the calibrator is coupled to the set. This should generate an audible beat note from the receiver's headphones or speaker, and TC1 is adjusted to produce the lowest possible beat note. It is not essential to do this, and a more than adequate accuracy will almost certainly be obtained if TC1 is left out, and the crystal is connected between the output and non-inverting input of the I.C.

It can sometimes be difficult to identify one harmonic from another, and using a crystal of a few MHZ in operating frequency will provide a S.W. signal of known frequency that will be of help here. Any signal of approximately known frequency can be used, however.

R.F. Generator

An R.F. signal generator covering the M.W. band can be a very useful aid to have in the workshop when aligning or servicing domestic radio equipment. The circuit of such a unit is shown in Figure 19, and this has a frequency coverage of approximately 1.6MHZ to 550kHZ.
It can also be used as a 470kHZ to 450kHZ I.F. alignment generator, as will be explained later.

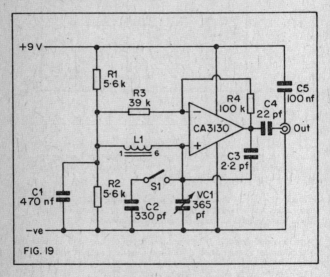

FIG. 19

This circuit is in some ways similar to the calibration generator, and like this previous circuit it is based on a non-inverting amplifier. An ordinary L-C tuned circuit is connected to the input circuit of the amplifier, and this consists of L1 and VC1. The latter is the tuning capacitor. A small amount of positive feedback is applied to the circuit by the inclusion of C3. This will only provide sufficient feedback for oscillation if there is a relatively high input impedance to the amplifier. The tuned circuit at the input has a high impedance at resonance, and a low impedance at other frequencies. Therefore the circuit oscillates at the resonant frequency of L1 and VC1.

When S1 is closed, the frequency range of the unit is reduced and lowered. It becomes approx. 600 to 400 kHZ, and then includes the usual broadcast receiver I.F.s of around 450 to 470kHZ.

Again, only a loose coupling between the oscillator and the radio is normally required. It is necessary to provide the unit with a reasonably accurate tuning dial if it is to be of maximum usefulness. It is an easy matter to calibrate the unit against the dial of any good quality M.W. radio of recent origin (the calibration of a receiver often deteriorates with age unless it is occasionally realigned).

When using the unit for I.F. alignment, close S1 and loosely couple the output of the oscillator to the primary of the first I.F. transformer. VC1 is then adjusted to produce maximum output at the receivers detector, after which the cores of the I.F.T.s are peaked.

It is not possible to make adjustments for maximum audio output, and since most receivers incorporate an A.G.C. circuit, this is not a very practical method anyway. If a multimeter is available this can be connected across the volume control of the receiver, positive test lead to earth for a positive earth receiver, and negative test lead to chassis in the case of a negative earth one. With the multimeter switched to **any** l/w volts range, the adjustments are then made for minimum output of the meter.

Another and even more simple method is to make adjustments for minimum background noise from the set. The volume must be turned well up so that there is plenty of background hiss.

The unit should work using M.W. tuned winding with the appropriate tuning capacitance, but the circuit has only been checked using a Denco Yellow Range 2T coil and a Jackson Type 0 365pf tuning capacitor. The necessary connection data for the coil is supplied with each one. Note that there are actually three windings on the coil, but only the one between pins 1 and 6 is used in this particular application. The other two are simply ignored.

A.F./R.F. Signal Injector

This really just consists of an A.F. squarewave generator which operates at about 1KHZ. The output is rich in harmonics extending well into the R.F. spectrum, and so this type of oscillator can be used for A.F., I.F., and R.F. tests.

The circuit diagram of this device is shown in Figure 20. The CA3130 is used as a non-inverting amplifier having a gain of about 12 times. Positive feedback is provided between the output and non-inverting input by C2. In order to obtain oscillation a gain of only unity is required, and so the circuit oscillates violently generating a squarewave output. Such a waveshape is rich in the desired harmonics. The values of

C2 and R3 are chosen to obtain a fundamental frequency in the middle of the audio frequency range.

As the peak to peak output of the oscillator is virtually equal to the supply voltage, the output of the circuit is excessive for many purposes. Therefore two outputs are provided, a high level one which is taken from the output of the I.C. via C3, and a low level one which is derived from the first output. This has series resistor R6 to reduce the output level, and capacitor C4 provides frequency compensation.

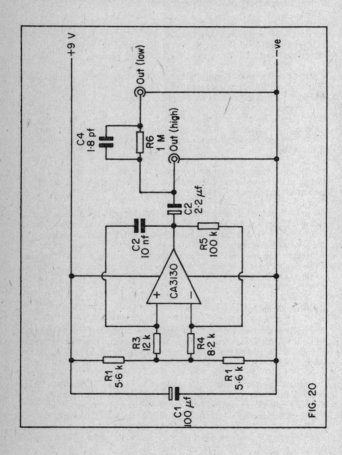

FIG. 20

Devices of this type are often built into large test probes as self contained units, complete with integral battery supply. Suitable containers are available for this purpose. This circuit could easily be miniaturised in this way, but it will be necessary to have a switch to enable the desired output to be selected, or to settle for a version having only a high level output.

This method is convenient to use, but it is not the only possible approach. It is quite in order to use a more conventional form of construction.

With this type of tester the first check is made at the speaker, and then one works back through the equipment injecting a signal at the input and output of each stage. In the case of transistorised designs this is usually at bases and collectors of the transistors.

When the audio signal is not heard at the output and a test is being carried out, the area in which the fault lies has been found. It is somewhere between the parts of the circuit to which the last test and the penultimate one were made. It is also possible that the fault is in the last stage to be tested, but this is not likely.

Thus if one injects a signal at the base of the driver transistor of an audio amplifier, and then at the collector of the preamplifier transistor, if the first test produces an audio output and the second one does not, then the fault probably lies in the coupling between the two stages. This suggests that the coupling capacitor between the two stages is faulty, or is not connected into circuit reliably. It is also possible that a fault in the preamplifier is causing the output of the injector to be short circuited, and is therefore preventing an output from being produced. This is less likely to be the trouble though, and should only be investigated if the coupling capacitor provides to be alright.

It is not practical to give a more detailed account of how this unit is used, but the above should give a fairly good idea of the way in which it is employed. It is really just a matter of using common sense and a logical approach.

It must be pointed out that it is not safe to use this tester on equipment which is mains operated and does not incorporate a mains isolating transformer. A lot of commercially produced domestic equipment falls into this category, and equipment of this type has its chassis connected to one side of the mains. Using this tester or any other piece of test gear (including other items described in this book) on this sort of equipment could easily lead to the user getting a powerful electric shock.

If in any doubt about this at all, do not try to fault find on mains powered equipment, and restrict such activities to battery power circuits until more experience and know how has been obtained.

A.F./R.F. Signal Tracer

In use this is rather like the opposite of a signal injector. When fault finding on a record player, for example, the first test would be made at the leads from the pickup. From here one would progress forwards through the amplifier until a signal can no longer be traced. Then, as before, the fault lies between the parts of the circuit which were subjected to the last two tests, or it could possibly be in the last stage that was tested.

Figure 21 shows the circuit diagram of the signal trace. This consists of a high gain audio amplifier preceeded by a demodulator circuit. The output of the unit is fed to a crystal aerpiece.

The CA3130 is used as an inverting amplifier having a voltage gain of 1000 times (60 dB.), and such a high gain is needed as the unit must provide a reasonable output from very weak signals, such as those found in the early stages of a radio receiver. D1, R3 and C5 are the demodulation circuit, and C3 provides D.C. blocking at the input.

From a look at the circuit diagram one would probably think that this circuit would not respond properly to an audio input, but in fact, this is not the case. An audio input signal will undergo a degree of distortion, and will be considerably attenuated, but this is not severe enough in either case to preclude the unit from A.F. testing. The level of distortion introduced by the demodulator is quite small, and the gain for audio tests needs to be reduced anyway as higher signal levels are found in audio stages than in R.F. and I.F. ones.

This type of equipment is another example of a kind of circuit that is often constructed as a self contained unit in a large test probe. Again it is not essential to do this, but if the unit is built into a normal case, the demodulator circuitry (C3, D1, R3, and C5) must be built into a probe.

The reason for this is quite simple. In order to avoid excessive stay pick up it is necessary to use a screened input lead. By having the demodulator in the test probe, the signal in the screened lead is at A.F. If the demodulator were to be constructed in the main unit, when R.F. were being carried out the screened lead would be carrying an R.F. signal. This is not a good idea as the capacity of the lead will heavily load an R.F. circuit to which it is connected, and it will also tend to have a detuning effect on any tuned circuit to which it is coupled.

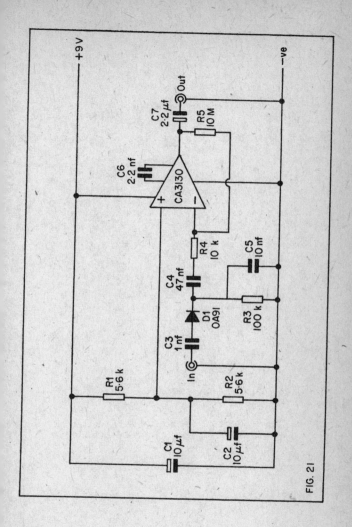

FIG. 21

Chapter 4
TEST EQUIPMENT

High Impedance Voltmeter

Test equipment is perhaps the area of electronics in which op. amps. are of most use to the amateur user. An operational amplifier is designed for use in D.C. amplifiers, and the most obvious use for one in this branch of electronics is in a high impedance voltmeter circuit.

A high impedance voltmeter is a multi-range voltmeter that is designed to overcome the main shortcoming of an ordinary multimeter. This is its relatively low input resistance, this usually being something in the order of 20k/V. for a good quality instrument. The current to move the needle of the meter must be provided by the circuit being tested, and this limits the input impedance of an ordinary multimeter if a reasonably robust meter movement is to be used.

Some electronic circuits operate at quite high impedances, and simply cannot provide the necessary current to operate the meter. For instance, the current in the base circuit of typical low level transistor stage is only a couple of micro-amps. The current needed to provide f.s.d. of a good quality multimeter is some 50 micro-amps. Trying to accurately measure the base potential of a low level transistor stage using an ordinary multimeter would obviously be pointless.

Most multimeters do not have very low voltage ranges, as the input resistance on such ranges would be unsatisfactory for almost all low voltage measurements.

By using an amplifier to drive the meter it is possible to produce a voltmeter requiring only a very small input current. Such an instrument can be used for virtually any voltage measurements without misleading readings being obtained. Low voltage ranges also become a practical proposition.

Figure 22 shows the circuit diagram of a simple high impedance voltmeter using a CA3130 I.C. The I.C. is used as a non-inverting D.C. amplifier having a voltage gain of eleven times. The output of the I.C. is used to drive a voltmeter circuit using R7 and a 100 micro-amp. meter. R7 is adjusted so that an input potential of 100mV. is required to produce f.s.d. of the meter.

Rather than use a single resistor to bias the non-inverting input of the I.C., a series of resistors in the form of a simple attenuator are used. This provides two additional input sensitivities of 1V. and 10V. f.s.d. It is necessary to use close tolerance resistors in the attenuator if high accuracy on all ranges is to be attained. In the case of R2, R3 and R4 this does not pose any problems, since 1 and 2% tolerance resistors are readily available in these values. Unfortunately, this is not the case when it comes to R1, as 10Meg. resistors would only seem to be readily available with a tolerance of 10%, which is too high. This will probably, therefore, have to be made up by adding a number of resistors in series. Ten 1 Meg. 5% tolerance resistors should provide adequate accuracy.

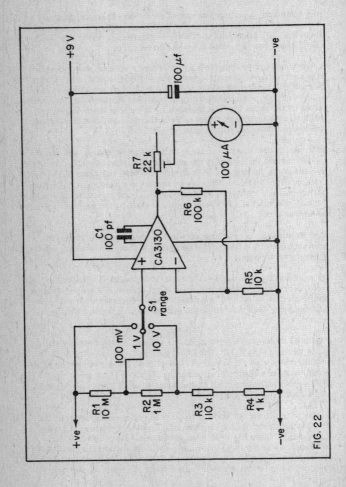

FIG. 22

It is an interesting feature of this circuit that it does not need to be powered from a dual balanced power supply. It operates perfectly well from a single supply of 9 volts. The reason for this is that the CA3130 will provide an output that can swing to within a few millivolts of either supply rail, provided the output is only lightly loaded, as it is here. Thus under quiescent conditions the voltage across the meter is insignificant, and it will read zero.

If most other op. amp. I.C.s were to be used in this circuit the meter would read beyond f.s.d. under quiescent conditions, as their outputs are incapable of swinging within a couple of volts of either supply line.

This circuit is rather vulnerable to pick up of mains hum and similar sources of interference, and it is necessary to house the instrument in a metal case to provide screening for the circuit. Screened test leads should also be used.

Adjusting R7 is quite a simple task. The only items needed for calibration are a multimeter and a 9 volt battery. Use the multimeter to measure the exact voltage of the battery, and then turn on the high impedance voltmeter and switch it to the 10V. range. Set R7 for maximum resistance, and then connect the test prods to the battery. Adjust R7 to produce the same voltage reading on the high impedance voltmeter as that just obtained on the multimeter. The unit is then ready to be used.

Resistance Meter

Most multimeters incorporate at least a couple of resistance ranges, but they are not usually as accurate as these ranges as they are on the voltage and current ones. This is mainly due to the fact that a very wide range of values are covered on each range, and that the resistance scale is logarithmic. This tends to make it extremely cramped at the high value end, and a small pointer error can produce wildly inaccurate readings. Another drawback is that the resistance scale reads in reverse, with the zero on the right hand side.

It is quite simple to produce a resistance meter having a linear forward reading scale by using the CA3130 as a high impedance voltmeter, to measure the voltage produced across the test resistor which is fed from a constant current generator. The circuit diagram of such a circuit is shown in Figure 23.

The theory behind the operation of this circuit is quite straight forward. Assume for example, that the constant current generator has an output of 1MA, and the voltmeter reads 0 to 10V. If a 10k resistor is connected into circuit, the meter will read f.s.d., as from Ohms Law it will be apparent that 10V. will be developed across the test resistor. If a 4.7k is placed in circuit a reading of 4.7V. will be obtained. For a 1k test resistor a reading of 1V. will be obtained. Thus, by calibrating the meter

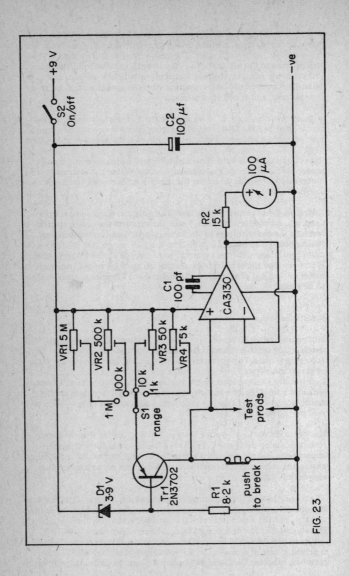

FIG. 23

in k ohms rather than in volts, a linear forward reading resistance meter is produced.

In the practical circuit a 2N3702 transistor (Tr1) is used as the basis of the constant current generator, and this can actually produce four different currents, the desired current being selected by S1. The use of four currents provides four resistance ranges, and these are 1k, 10k, 100k, and 1M f.s.d.

The I.C. is used as a unity gain amplifier and it drives a conventional voltmeter circuit using R2 and a 100 micro-amp. meter. It is necessary to use a high impedance voltmeter, as it is obviously important that no significant current is drained from the constant current generator. On the higher ranges there is insufficient current in the test resistor to operate a moving coil meter anyway.

With no test resistance in circuit the meter would be taken beyond f.s.d. if the push button switch was not included in the circuit. This short circuits the test leads and keeps the meter at zero until the test resistor has been connected. Then this switch is operated so that a reading may be taken.

The unit is calibrated against four close tolerance (1 or 2%) resistors. One is needed for each range, and each resistor should have a value equal to, or a little less than the f.s.d. value of the range is to be used to calibrate. For instance, an 820 ohm resistor could be used to calibrate the 0 to 1k range.

With the unit switched on and set to the 1k range, the test resistor is connected across the test leads. Set all the presets for maximum resistance and operate the push button switch. Adjust VR4 for a reading of 82 on the meter, so what ever value is appropriate for the test resistance used.

A similar process is then used for the calibration of the other three ranges.

Sinewave Generator

Operational amplifiers make the ideal basis for a number of types of waveform generators. This simple sinewave generator can be used as a very simple A.F. signal generator and it provides an output of 7.5HZ to 75kHZ in four ranges. The coverage of each range is as follows:—
7.5 to 75HZ, 75 to 750HZ, 750HZ to 7.5kHZ, and 7.5 to 75kHZ.
The circuit diagram of the unit is shown in Figure 24.

This is based on the well known Wien Bridge type of oscillator. The resistive parts of the R-C network are R3, R5, and VR1, VR1 is a dual gang linear carbon potentiometer, and this enables the time constant of the Wien network to be varied, and it acts as the frequency control.

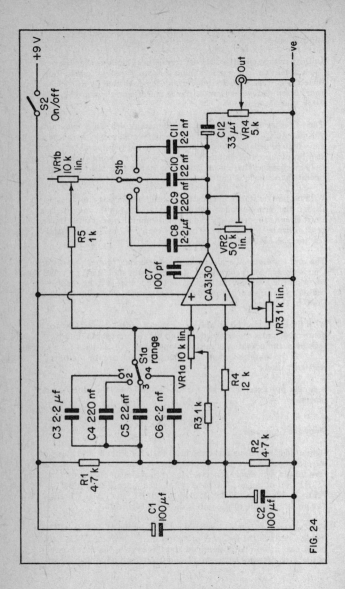

FIG. 24

The capacitive elements of the Wien network are formed by the two capacitors that are switched into circuit by S1. This provides four sets of capacitors, and it is this that gives the unit its four frequency ranges.

The role of the CA3130 in this circuit is quite simple. It must provide a non-inverting amplifier having a very high input impedance and a voltage gain of only approx. three times. It may at first appear that the voltage gain could be set at three by using fixed resistors in the feedback loop, but in practice this would not be satisfactory. The gain of the circuit must be very accurately adjusted, as if it is only fractionally low the circuit will not oscillate, and if it is a little on the high side the output waveform will be seriously distorted. The Wien network provides the frequency selective positive feedback path between the output and the non-inverting input of the I.C., which is necessary to cause the circuit to oscillate. The voltage gain of the circuit is manually adjusted to the appropriate level using VR3.

Before the unit is ready for use it is necessary to adjust VR2. Connect the output of the unit to a crystal earpiece, or an amplifier and speaker combination. Set VR1 and VR2 for about half maximum resistance, and then turn the unit on. Switch S1 to range 2 and set VR2 for almost maximum resistance. This should cause a tone to be heard from the earpiece or speaker. If VR2 is now adjusted for a lower resistance, the tone will rise in pitch, and should have a more pure sound. Adjusting it for an even lower resistance will cause oscillation to cease.

VR1 is adjusted to the position that corresponds to the lowest resistance that produces oscillation. In use, VR3 is similarly adjusted, and it will need slight readjustment each time the setting of VR1 and S1 altered.

VR4 is the output level control.

A.G.C. Sinewave Generator

It is obviously rather time consuming having to manually adjust the gain of the amplifier to the sinewave generator just described. Most signal generators of this type incorporate an A.G.C. circuit which automatically adjusts the circuit gain to the proper level. Figure 25 shows the circuit diagram of such a unit.

This is identical to the previous circuit except for the inclusion of the A.G.C. circuitry. The A.G.C. action is provided by Tr1 which is a field effect transistor. When the unit is initially switched on, the gate of TR1 is connected to the negative supply by way of R7. This strongly reverse biases the f.e.t. which in consequence exhibits a very high d. to s. resistance. This gives the amplifier a high gain and the circuit starts to oscillate.

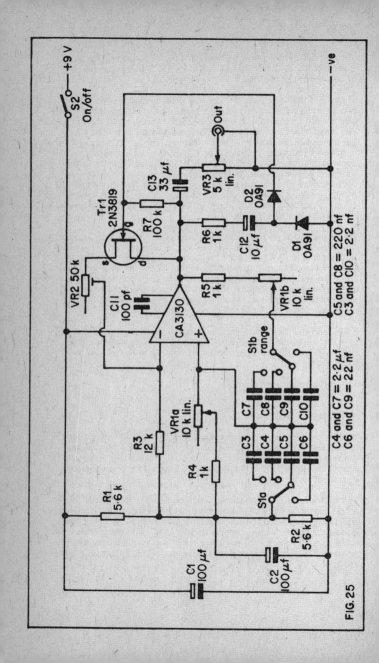

FIG. 25

The output of the oscillator is fed to a rectifier network via Rc and C12. The positive output pulses of D1 and D2 are fed to the gate of Tr1 where they have the effect of decreasing its reverse bias and, therefore, also the gain of the amplifier. The greater the amplitude of the oscillations, the greater the decrease in the amplifiers gain. This circuit action thus stabilises the gain of the circuit at the correct level. VR2 enables the circuit to be adjusted so that the A.G.C. action provides optium results.

To enable VR2 to be adjusted the unit must be coupled to a crystal earpiece or an amplifier and speaker combination. Switch the unit to range 2 and set VR1 for maximum frequency. With VR2 set at or towards either extreme, the output note of the unit will sound rather harsh. Somewhere towards the centre of its adjustment range it should be possible to find a range of settings that give a very pure sounding note. VR2 is set at the centre of this range of settings.

An alternative method is to connect the unit to an A.C. millivoltmeter or a multimeter that has a low A.C. voltage range. VR2 is then adjusted to produce an output level of 500mV. R.M.S.

Calibration

It is usual to calibrate an audio generator in terms of output frequency, and it is a simple matter to mark a suitable dial around the control knob of VR1. If a calibrated audio generator can be borrowed, it is possible to calibrate these units against this. This can either be done using oscilloscope techniques (if a suitable scope is available), or it can be done using an aural technique. The aural technique is the one most constructors will have to use, and this is really very simple.

The output of each generator must be connected to an earpiece, or something of this nature, that enables the output notes of the generators to be heard. The calibrated generator is then set to, say, 100HZ, and the other generator is then adjusted to produce the same note. It is then operating at 100HZ and its dial can be marked accordingly at the appropriate point. The same procedure is then used to calibrate the dial at other points, until a fully calibrated dial is produced.

If a suitable signal generator is not available, an alternative is to use a musical instrument and a piano scale frequency chart to provide the calibration frequencies.

Sine To Squarewave Converter

For some audio tests a squarewave rather than a sinewave is required. Many signal generators are therefore designed to provide both types of output. The usual way of achieving this is to use a sinewave generator to produce the sine output, and then use this to drive a squaring circuit to provide a square output.

A suitable squaring circuit for use with the sinewave generators just described is shown in Figure 26. This is also suitable for use with any other sinewave generator having an output amplitude in the range 100mV. to 3V R.M.S.

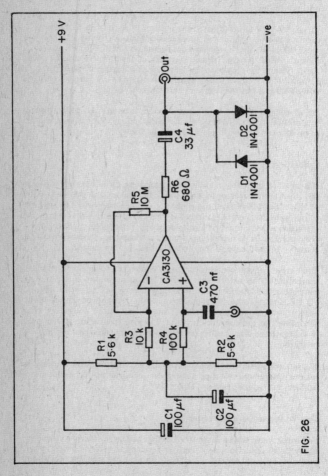

FIG. 26

The CA3130 is used as a non-inverting amplifier with a voltage gain of about 1000 times. Obviously this will not produce an output of 1000V. from an input of 1V., but will clip the signal at the output to produce a squarewave of reasonable quality.

The output of the I.C. will have a peak to peak voltage swing of about 9V., which is more than is normally required. R6, D1, and D2 are used to clip the output of the I.C. at about 1V. peak to peak. Apart from decreasing the output, this also improves the output waveform slightly.

The unit is suitable for use with input frequencies from a few HZ to a few tens of KHZ. Outside these limits the output waveform will greatly deteriorate.

Squarewave Generator

If separate sine and squarewave generators are preferred, or only a squarewave generator is required, the circuit shown in Figure 27 can be used. This has a frequency coverage of 5HZ to 50kHZ in four ranges. There coverage of each range is as follows:— Range 1, 5 to 50HZ; Range 2, 50HZ to 500HZ; Range 3, 500HZ to 5kHZ; Range 4, 5 to 50kHZ.

Here, once again, the CA3130 is used as a high gain non-inverting amplifier. Positive feedback is provided between the output and non-inverting input of the I.C. by VR1 and whichever of the set of four capacitors is switched into circuit by S1. VR1 varies the time constant of the feedback network, and is the frequency control. S1 is the range switch, with a different feedback capacitor being used for each range. As in the case of the previous circuit, a simple clipping circuit is included at the output to reduce the output to about 1V. peak to peak, and to provide an improved waveform.

Low Current P.S.U.

Often in electronics it is necessary to power a piece of equipment from a well stabilised supply. Whether the equipment is battery powered or uses a mains power pack, a high quality voltage regulator circuit is needed. The circuit shown in Figure 28 can be used to provide a well regulated output of between about 4 volts and 9 volts. It is only intended for low output current applications, and the maximum output current that should be drawn from a CA3130 is 20mA. The input voltage should be at least one volt higher than the required output voltage, and the input potential should never be in excess of 16 volts.

The circuit operates in the following manner. Assume that the slider of VR1 is at the top of its track, the I.C. then has 100% negative feedback, and the output voltage will be equal to that at the non-inverting input of the device, or about 3.9V. in other words.

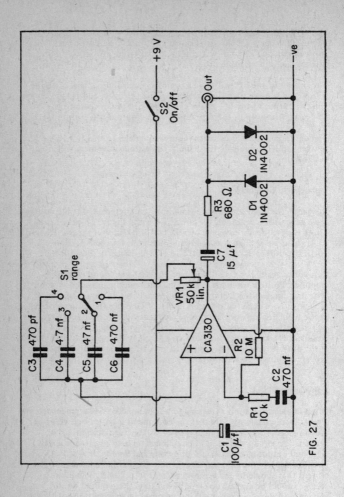

FIG. 27

If the slider of VR1 is taken down towards the bottom of its track, about half the output voltage will appear at the inverting input of the I.C. The circuit will try to maintain the two inputs of the I.C. at the same potential, and so the output will go to about double the voltage at the non-inverting input, or about 8 volts in actual fact. With the slider of VR1 at the very bottom of its track, a little over 9 volts is required at the output to maintain the balance of the two inputs.

Thus VR1 can be used to vary the output voltage over a range of about 3.9 to 9 volts.

If for some reason the output voltage of the I.C. should fall, because of increased loading on the output for example, the inputs will be unbalanced with the non-inverting input being higher in potential than the inverting one. This causes the output to go more positive in order to restore the balance of the inputs, and in doing so it returns the output potential to its original level. The output voltage of the circuit is thus very well stabilised, and any significant variations in the output voltage are due to the limitations of the zener reference voltage, and not those of the CA3130.

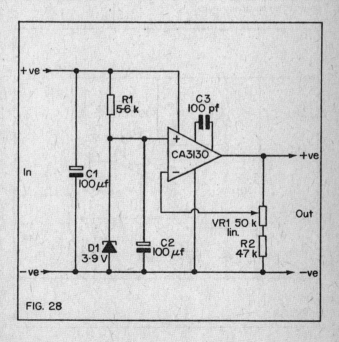

FIG. 28

High Current Version

It is a simple matter to modify this circuit to provide a higher output current, and all that is needed is an emitter follower buffer stage at the output. The circuit diagram of the high current version of the regulator is shown in Figure 29. This is capable of handling output currents of up to about 500mA.

The circuit operates in exactly the same manner as the original circuit. The feedback loop is connected in the emitter circuit of Tr1 rather than in the output circuit of the I.C., as in this way the feedback compensates for any variations in the voltage dropped across Tr1. This gives improved regulation over the alternative method.

In both these regulator circuits, VR1 can be a preset resistor if the unit is only going to be used to supply a single output voltage.

Tr1 must be fitted with an adequate heatsink as otherwise it will overheat and be destroyed. An area of about 200 sq. c.m. of 16 s.w.g. aluminium should be sufficient provided the unit is not used beyond its specified maximum limits.

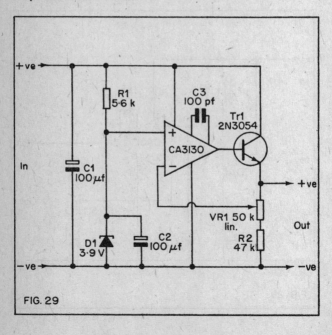

FIG. 29

Continuity Tester

Although a continuity tester is one of the most simple items of test gear, it is nevertheless of the most useful to have around the workshop. This unit consists of a simple audio oscillator which feeds a loudspeaker. The two test prods are connected in the supply lines, and when they are short circuited, or connected together through a fairly low impedance, power is applied to the circuit and an audio tone is produced.

This is more convenient than using the common alternative of a multimeter switched to an ohms range. Continuity testing with devices of this nature, where there is a visual indication of continuity, is often inconvenient as it is necessary to take ones eyes off the test prods in order to ascertain whether or not there is continuity. This circuit provides an audible indication, and there is therefore no need to remove ones attention from the test prods.

The circuit diagram of the unit appears in Figure 30. This consists of a D.C. amplifier which directly drives a high impedance loudspeaker. The speaker must have an impedance of 75 or 80 ohms if the units is to function well. Positive feedback is needed in order to make the circuit oscillate, and this is provided by C3.

The output across the speaker is a rapid stream of brief pulses, and this produces a rather rough but clearly noticeable output tone of about 500HZ in frequency.

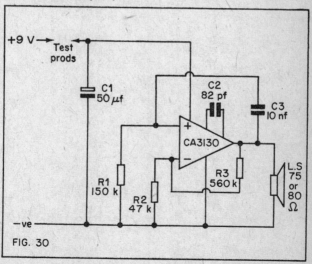

FIG. 30

Chapter 5

HOUSEHOLD PROJECTS

Metronome

A conventional metronome uses a purely mechanical mechanism to produce a series of clicks at regular intervals. It is quite easy to simulate this electronically, and the simple circuit shown in Figure 31 performs this task.

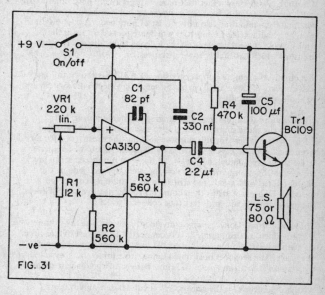

FIG. 31

This circuit is based on the continuity tester circuit just described. However, the circuit values have been adjusted to provide a much slower rate of oscillation, and by enabling the time constant of the feedback circuitry to be varied, the oscillation frequency is made variable. VR1 can be adjusted to produce any beat rate from about 50 beats per minute to over 200 beats per minute.

In this particular application the output drive of the CA3130 is not sufficient to produce an adequate volume from the speaker without some additional amplification at the output. An emitter follower buffer stage has therefore been included at the output, and this uses only two components (Tr1 and R4).

It is necessary to mark a dial around the control knob of VR1 so that the unit can be quickly set to any desired beat rate. This is quite easily done as the relatively low frequency range of the unit means that it is quite possible to count the number of pulses produced per minute. It will be quicker if one counts the number of pulses emitted during a fifteen second period, and then multiplies this by four to find the number produced per minute.

Rain Alarm

It is quite a well known fact that pure water is a very poor conductor, and it would probably be more accurate to call it an insulator. Fortunately, rain drops do not consist of pure water, and contain relatively high levels of impurities which are picked up from the atmosphere. These dissolve in the rain drops to produce very weak solutions which have fairly low resistances.

The circuit of a simple rain alarm using a CA3130 is shown in Figure 32. This consists of three basic parts, the sensor, an electronic switch, and an audio alarm circuit.

There is more than one way of arranging a suitable sensor, but probably the most simple method is to use a piece of stripboard. 0.1 in. matrix is best for this purpose as it has the most strips for any given area. A piece having 24 strips by 50 holes should be adequate. If the strips are numbered 1 to 24, all the even numbered strips are connected together by link wires on the non-coppered side of the board. All the odd numbered strips are then similarly connected together. One set of strips then connects to the positive supply line of the rain alarm circuit, and the other connects to R6.

The sensor is positioned at any convenient spot outside the house where it is not shielded from rainfall. It is connected to the rest of the circuit via twin insulated cable, and this cable can be several yards long if necessary. The sensor is positioned copper side up so that any raindrops that fall on it form an electrical bridge between two adjacent strips.

With no raindrops on the sensor, Tr1 is cut off and only minute leakage currents will flow in the circuit. This is very important as it ensures that the battery has a very long life, and is not run down even when the alarm is not sounding.

When water is present on the sensor, a base current is supplied to the unit through R6 and the senor. R6 is a current limiting resistor, and is needed to ensure that Tr1 does not pass an excessive base current. Tr1 is used as the electronic switch, and when it is biased into conduction it supplies power to a simple audio oscillator utilising a CA3130 I.C. This causes an audio tone to be emitted from the unit which is, of course, situated inside the house where it will alert the user.

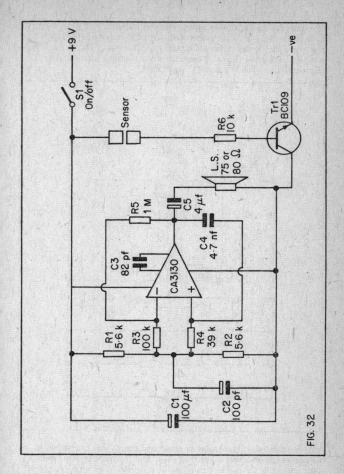

FIG. 32

Light Switches

Switches that are operated automatically by changes in light intensity are among the most useful and popular of electronic projects. They can be used as the basis of many gadgets, such as porch lights that automatically turn on at night, and off at daybreak. Burglar alarm systems can also incorporate this type of switch. They can also be used in applications outside the home, such as in automatic parking lights on cars.

When used as a comparator, an operational amplifier makes an ideal basis for a photo sensitive switch. The circuit diagram of a simple photo switch incorporating a CA3130 I.C. is shown in Figure 33. This is designed to close the relay contacts when the light falling on the photocell drops below a certain preset level.

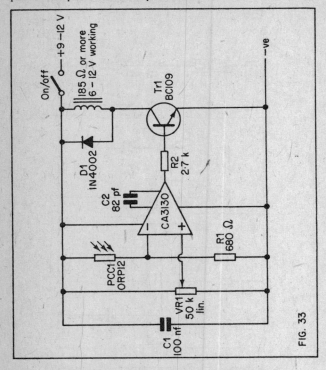

FIG. 33

VR1 is adjusted so that under normal conditions there is a higher voltage at the inverting input than there is at the non-inverting input. This causes the output of the I.C. to be normally low, with Tr1 cut off and no significant current flowing through the relay coil.

If the level of light falling upon PCC1 should now drop for some reason, the resistance of PCC1 will increase and the voltage at the inverting input will fall. If it falls below the voltage at the non-inverting input, the output of the I.C. will swing high and will turn Tr1 hard on. This will cause the relay to be activated.

When the light level on PCC1 returns to normal, the relay will turn off once again. By adjusting VR1, this circuit can be adjusted to switch over at virtually any light intensity one desires.

If the relay has change over contacts, it can be connected so that it either switches the ancillary equipment on when the light level falls below the threshold level, or so that it switches it on when the light level rises above the threshold level. If the relay only has make contacts, it can only be used to perform the former.

The circuit can be modified very easily to enable a relay having only make contacts to turn the ancillary equipment on when the light intensity rises above the threshold level. The modified circuit is shown in Figure 34.

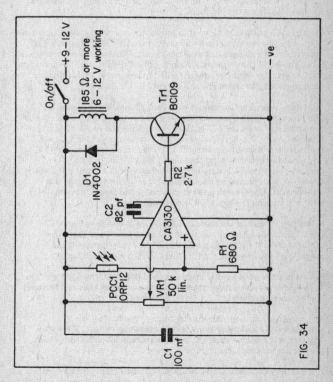

FIG. 34

All that has been done here is that the inputs of the I.C. have been swopped over. Now, under normal conditions VR1 is adjusted so that the voltage at the inverting input is more than that at the non-inverting input, just as before. However, when the light intensity on the photo conductive cell increases, the voltage at the non-inverting input increases, and the output of the I.C. goes high. This operates the relay, and also the ancillary equipment. This circuit thus operates in the reverse manner to that of the previous design.

Latching Circuits

For certain applications a photo switch that latches is required. This type of switch differs from those just covered in that once the relay comes on, it remains on, regardless of any further changes in the light intensity falling on the photocell.

Circuits such as this can be used in burglar alarms, for example. A photo cell can be positioned so that when light from an intruders torch falls upon it, the circuit operates an alarm circuit. Obviously this arrangement is of little use if the alarm only sounds while the light from the intruders torch is actually on the photocell. What is required is a circuit where once triggered, the alarm remains on until it is turned off.

A latching version of the circuit of Figure 33 appears in Figure 35. This works in much the same manner as the original circuit, except that when the output of the I.C. goes high, an additional transistor (Tr1) is turned on by the current flowing from the output of the I.C. and through current limiting resistor R2. Tr1 is turned hard on, and the voltage at the inverting input becomes almost equal to that of the negative supply rail. Changes in the resistance of PCC1 will not greatly affect this voltage, and so the circuit latches in this state until it is switched off. Upon turning the circuit on, it will function normally until it is triggered, whereupon it will latch again.

A latching version of the circuit of Figure 34 is shown in Figure 36. Once again, this operates exactly in the same manner as the original until the output of the I.C. goes high. Then Tr1 is turned on and the potential at the inverting input of the I.C. falls to virtually zero.

Even when the photo cell is in almost total darkness, the voltage at the non-inverting input is more than that at the inverting one, and the relay contacts remain closed.

The relay used in these circuits can be any type having a coil resistance of 185 ohms or more and an operating voltage of 6 to 9 volts. The contacts must be chosen to suit the particular application which the circuits are employed.

It is not very economical to power these circuits from ordinary batteries as this type of device is normally left turned on for prolonged periods.

Also, the current consumption is quite high when the relay is closed, being something like 10 to 30mA., depending upon the type of relay used. It is therefore advisable to either power these circuits from a mains supply or rechargeable batteries, whichever is most appropriate to their application.

The latching circuits can be reset by simply turning the units off, and the then switching them on again. If a separate reset switch is preferred, this can be provided by connecting a push to make non-locking push button switch between the negative supply rail and the base of Tr1. This modification is suitable for use with either circuit.

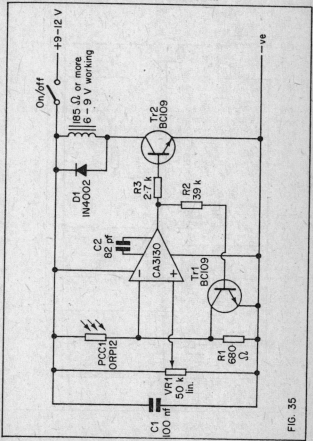

FIG. 35

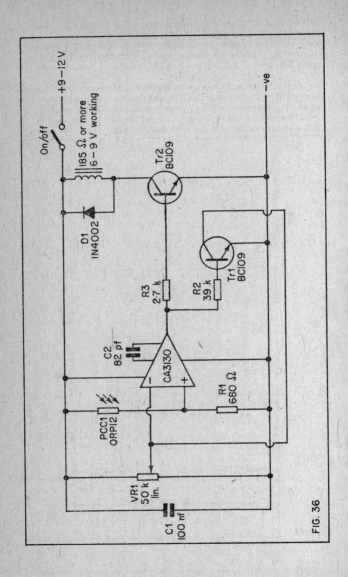

FIG. 36

Sound Activated Switch

Sound activated switches have a multitude of uses. In the home, the most obvious use for one is as a baby alarm. They can also be used in burglar alarm systems. They also find uses in the field of amateur radio (in VOX, or voice operated switch systems), and can be used to automatically operate a tape recorder, when dictating something for example.

Operational amplifiers can easily achieve the high voltage gains required in this type of equipment, and can be used as the basis of a very simple but effective sound activated switch. Figure 37 shows the circuit diagram of such a unit.

The purpose of the I.C. is to amplify the very low level microphone signals to bring them up to a level which can be used to operate a switching transistor. In this design the microphone is actually a miniature high impedance loudspeaker which is used as a sort of moving coil microphone. The I.C. is used in the inverting mode and it has a voltage gain of more than 10,00 times. Even an input signal of less than a millivolt generates an output of a few volts peak to peak at the output of the I.C.

The output of the I.C. is fed via C5 to a rectifier and smoothing network using D1, D2, and C6. The output of this network is a positive D.C. bias. Provided a reasonably high sound level is received at the microphone, this bias will be strong enough to bias Tr1 virtually into saturation. This operates the relay which, in turn, switches on the controlled equipment.

This circuit has hysteresis, which is desirable in most applications. The hold on time of the circuit with the values specified is about 1 to 2 seconds. If required, this can be altered by changing the value of C6.

In order to obtain good sensitivity and battery economy it is necessary to use a relay having a fairly high coil resistance. This should preferably be 400 ohms or more. The author used an RS open P.C. relay on the prototype. This has a coil resistance of 410 ohms and an operating voltage (nominal) of 6 volts. The relay should not have an operating voltage of less than 6 volts. If high speed operation is required, a reed relay should be used.

The prototype is quite sensitive, and talking at normal volume levels causes the unit to operate even at a distance of several feet. The exact sensitivity of each unit will vary according to the type and make of speaker/microphone used, current gain of Tr1, and similar parameters. The sensitivity obtained should always be high enough for the majority of applications though.

One point must be borne in mind when constructing this equipment. The relay and the microphone should not be housed in the same case, and should not even be in close proximity to one another. If they are,

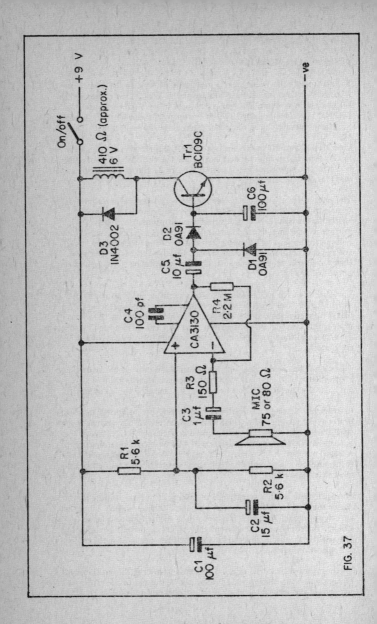

then as the relay turns off, the sound it produces will activate the unit. After a second or so the unit will turn off again, and the noise generated by the relay will again activate the unit. The circuit will continue to oscillate in this manner for as long as power is applied to the circuit.

In most applications there is no need to mount the microphone and the relay in the same casing anyway. If, for example, the unit is employed as a baby alarm, the microphone would be mounted in its own case near the baby with the rest of the unit in a separate case situated near the user. Screened cable must be used to connect the microphone to the main unit, and as a low impedance microphone is used, the cable can be several yards long if necessary.

Some readers may be puzzled about the inclusion of a diode across the relay coil in this circuit, and in the photo switches described previously. This is a protection diode which is needed because of the high reverse voltage that is developed across the relay coil when the supply is switched off. This voltage is generated as the magnetic lines of force around the solenoid quickly decay and cut through the turns on the coil. The voltage generated can be high enough to damage any of the semiconductor devices of the circuit, even though it is at a high impedance.

D3 acts as a sort of low voltage zener diode, and it limits this voltage to only about 0.5V. in peak amplitude. There is no need to add any form of current limiting circuit in series with D3, as this current is limited to a safe level by its high source impedance.

Do not be tempted to omit D3, as this would almost certainly turn out to be a false economy.

Latching Version

This is another example of a device that must be made to latch if it is to be usable in certain application, such as burglar systems. This is quite simple to achieve, and the modified circuit diagram for this purpose appears in Figure 38.

This circuit operates in exactly the same way as the original until the relay is energised. When this happens, Tr2 is turned on by the base current flowing via R5. This causes a current to flow through the emitter and collector of Tr2, through R6, and into the base of Tr1. Tr1 is held on by this current, and even if no further sound is received by the microphone, it will remain on.

Tr1 and Tr2 are, in fact, operating as a sort of thyristor.

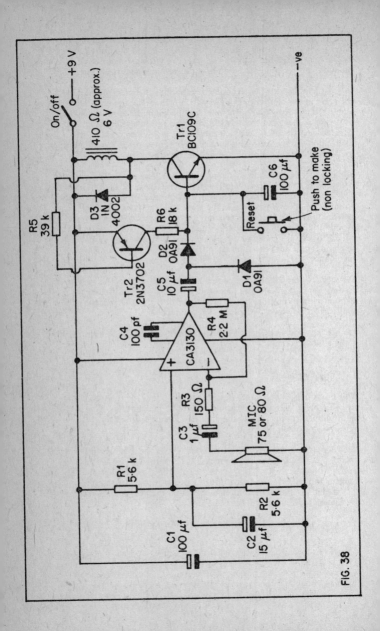
FIG. 38

Christmas Tree Lights Flasher

The usual way of getting the lights on a Christmas tree to flash on and off is to use a string of series connected bulbs, with one of these being of the bi-metal strip flashing type. When this bulb is on , it completes the whole circuit, and all the bulbs come on. When it is off, it breaks the series circuit, and all the bulbs go off.

This arrangement is very simple and works quite well, but it does have the slight drawback that an ordinary flashing bulb is rather erratic in operation, and does not usually give a very regular flash rate.

The circuit of Figure 39 can be used to operate the lights, and this will flash them on for a period of about 1 second with a similar time elapsing between flashes.

A squarewave generator utilising a CA3130 I.C. forms the basis of the circuit. Normally when a slow rate of oscillation is required, as it is here, a high value of timing capacitor must be used. This is not the case here, however, as the high input impedance of the I.C. enables a high value of timing resistor (R4) to be used, and so a relatively low timing capacitance (C3) can be used.

The output of the I.C. drives a common emitter amplifier via the current limiting resistor, R6. When the output of the I.C. goes high, Tr1 is turned on and the relay is energised. When the output of the I.C. is low, Tr1 is cut off and the relay receives no significant current. The relay contacts are used to control the lights, and it is essential to check that these have a high enough rating for the voltages and currents involved. It is advisable to have contacts that are rated well in excess of the current drawn by the lamps, as when power is first applied to a lamp a heavy surge current flows. This is because the cold resistance of a lamp is far lower than its normal hot working resistance. With the lamps being constantly turned on and off there is a constant string of current surges for the contacts to handle.

One might think that the relay would be short lived in this circuit anyway, as with such rapid switching it would soon wear out. This is not the case though, as any modern relay of reasonably good quality is guaranteed to last for several hundred thousand operations.

The unit can be battery operated, but since ti is controlling a mains load it would be more logical to construct a mains power supply for it.

Simple Organ

The field of electronic musical instrument and effects is one which has increased greatly in popularity over the past few years, and it must now rate as one of the most popular branches of electronics. Electronic musical instruments need not be complicated, and a simple electronic organ can be built using very few components indeed. A simple circuit of this type using a CA3130 I.C. as a tone generator is shown in Figure 40.

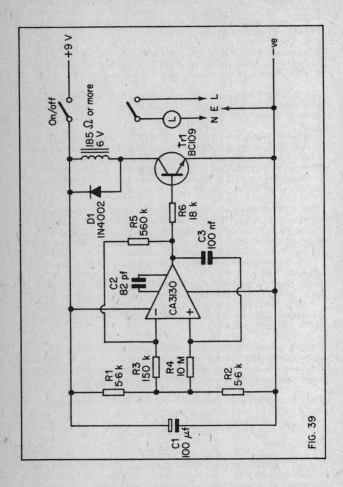

FIG. 39

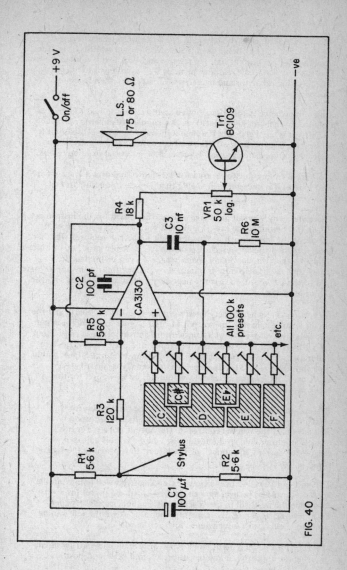

FIG. 40

Here the CA3130 is used once again as a squarewave generator, with the output of the I.C. being used to drive a common emitter transistor amplifier. R4 is a current limiting resistor and VR1 is the volume control. This is a very economical arrangement as D.C. coupling is used, and a D.C. blocking capacitor and bias resistor for Tr1 are therefore unnecessary.

A high impedance loudspeaker is used as the collector load for Tr1, and quite a high volume level (for such a simple instrument) is available. The circuit is most efficient if a high impedance speaker is used, but it will work using speakers with impedances as low as 15 ohms. When using a low impedance speaker R4 should be increased in value to 39k.

A different timing resistor is used for each note, and preset resistors are used here as it is necessary to be able to adjust each tone for tuning purposes.

There are many ways of arranging a simple keyboard for the instrument, but almost certainly the easiest and most practical method is to etch one along the lines shown in the circuit diagram. The enclosed shaded areas represent the coppered areas of the P.C.B. A test prod, or anything similar to this (wander plug, banana plug, etc.) can be used as the stylus, and the desired note is obtained by placing this on the appropriate part of the keyboard, so that the necessary circuit is completed and the circuit oscillates.

R6 ensures that when the stylus is not placed on the keyboard, and the circuit is not oscillating, the output of the I.C. goes low and Tr1 is cut off. This gives a very low quiescent current consumption of only about 1mA. If R6 were to be omitted, the output of the I.C. would go high under quiescent conditions, and a current of up to about 50mA. would flow through Tr1 and the speaker. Apart from giving poor battery economy, a large standing current would not be very good for the speaker.

The current consumption of the unit when the tone generator is oscillating varies from about 1mA. to 30mA., depending upon how well advanced the volume control is. The output stage is a sort of Class B amplifier, and so the higher the volume level is adjusted, the greater the current consumption. The circuit thus provides the longest possible battery life.

The unit can be tuned over a range extending from well below middle A to the A several octaves above this. By adjusting a preset for a very low value the unit will in fact oscillate at frequencies at the upper limit of human hearing. By increasing the value of C3 the unit can be made to oscillate at frequencies as low as one wishes.

It is therefore possible to obtain a range of several octaves if required, by using the appropriate number of presets and keyboard positions.

For most purposes a single or two octave version will be sufficient. This would have a compass from middle C to the C either one or two octaves above this. Thirteen 100k presets are needed for a single octave unit, and twenty five are required for a two octave version. This gives a chromatic scale.

The unit is easily tuned against a piano, organ, pitch pipes, or virtually any properly tuned musical instrument. A reasonably musical ear is required for this, as not everyone will find it possible to tune the notes on the organ to those produced by the instrument it is being tuned against. However, most people will find that this is not too difficult after a little practise.

It should perhaps be pointed out that this instrument is not polyphonic, and cannot be used to play chords. It can only produce one note at a time. The tone is quite pleasant though, particularly at the low frequency end of the compass, and if a reasonably large speaker is used. The absence of chords also makes the unit very easy for a beginner to play.

Chapter 6
MISCELLANEOUS PROJECTS

As will probably have become apparent by now, op. amps. in general, and the CA3130 in particular, are very versatile. They can be used to perform a number of diverse applications which do not fit into any of the categories so far covered.

In this chapter a number of interesting circuits of varying types will be described, all of which are built around a CA3130 device.

Touch Switch

Touch switches have only redently become widespread in use, and they do have definite advantages over more conventional mechanical switches in certain applications. However, it is probably their novelty value that largely accounts for their popularity.

Figure 41 shows the circuit diagram of a simple touch switch utilising a CA3130 I.C. When the supply is connected, the voltage at the inverting input is taken to half the supply potential as it is fed from the potential

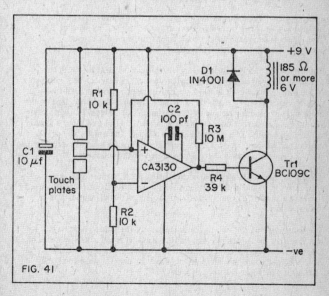

FIG. 41

divider network formed by R1 and R2. The non-inverting output is fed from the output of the I.C. via R3, and it will therefore remain at virtually earth potential. Thus at switch on the output of the I.C. is low, and Tr1 receives no significant base current. Tr1 is cut off and the relay is energised.

If a finger is now lightly touched onto the two upper touch plates, the resistance between these two plates will be something in the order of a few Meg. ohms, and the voltage at the non-inverting input will go above that present at the inverting input. This will cause the output of the I.C. to go high, Tr1 will turn on, and the relay will be energised.

Even if the finger is now removed from the touch plates, the relay will remain on. This is because the non-inverting is coupled to the output of the I.C. by R3, and it will therefore be held at a higher potential than the inverting input.

In order to set the circuit back in its original state it is necessary to touch the middle and lower touch plates. The operator's skin resistance will then bridge these two contacts, and R3 together with this skin resistance will form a potential divider. The skin resistance will be less than that of R3 and so there will be a lower voltage at the non-inverting input than there is at the inverting one. The output of the I.C. will go low and the relay will be de-energised.

The circuit will again latch in this state, as the non-inverting input will be held low as it is coupled to the output of the I.C. by way of R3. There circuit can, of course, be set to the on state once again by touching the upper touch contacts.

There must be many possible uses for this circuit, and it could, for example, be used as a touch operated wavechange switch in a radio set. If the unit is to be used in a simple on/off application, it might be possible to remove D1 and the relay, and connect the controlled equipment in the collector circuit of Tr1 instead. The controlled equipment should have a current consumption of 50mA. or less and need a supply voltage of 9 volts. The current consumption of the unit in the off state is about 250 micro-amps. The current consumption when it is in the on state is dependant upon the coil resistance of the relay used.

A suitable touch plate can be etched using p.c.b. techniques. An alternative is to use stripboard in a similar manner to the way it is used to make a sensor for the Rain Alarm project. A little ingenuity is required here.

The sensitivity of the circuit is controlled by the value given to R3. If necessary, the sensitivity can be increased by increasing the value of R3. Resistors having values of more than 10Meg. are not readily available, and it will be necessary to make up any increased value by adding two or more resistors in series.

Flashing Pilot Light

A problem with most small battery operated signal generators and similar pieces of equipment is that it is very easy to inadvertently leave them on. Such units rarely incorporate a pilot light as these consume a relatively high current. Even light emitting diodes need about 10 to 20mA. to make them glow brightly. Some types need even more than this.

It is possible to reduce the current consumed by a L.E.D. pilot lamp by pulsing it. If it is pulsed on for say, 1/3 of a second at one second intervals, the current consumption when averaged out will be only about one third of the value needed to keep the L.E.D. on all the time. Furthermore, a flashing pilot light is more noticeable than a conventional one anyway.

When the output of a CA3130 I.C. is in the low state it has a current consumption of only about 200 to 300 micro-amps. (from a 9V. supply), and so it makes an ideal basis for this type of circuit. The circuit diagram of a flashing pilot light using a CA3130 I.C. is shown in Figure 42.

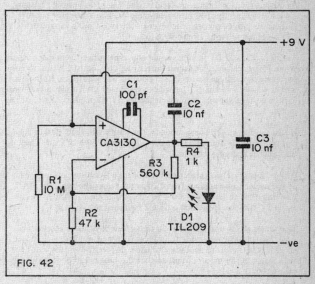

FIG. 42

This circuit is very similar to those of the Continuity Tester and
Metronome which appear in Figures 30 and 31 respectively, in this
book. It is a non-inverting D.C. amplifier with positive feedback
between its output and input. It briefly switches the L.E.D. on at
approximately one second intervals. The average current consumption
of the unit is only about 2.5mA.

Usually this type of circuit employs large electrolytic capacitors as a
high value timing capacitor is needed to provide the low oscillation
frequency. The high input impedance of the CA3130 enables the low
oscillation frequency to be obtained using a timing capacitance of only
10nf. This unit is therefore easily miniaturised.

Analogue Stopclock

This simple circuit acts as a two range analogue stopclock have f.s.d.
values of 10 and 100 seconds. The circuit diagram of this device
appears in Figure 43.

The CA3130 is used as a non-inverting unity gain buffer amplifier, and
its output feeds a voltmeter circuit using one of the preset resistors,
R4 and R5, and the 100 micro-amp. meter. The voltmeter reads the
voltage developed across whichever of the two timing capacitors (C1 or
C2) is connected into circuit by S1a.

Tr1, R1, R2, and D1 form a simple constant current generator circuit.
When PB1 is closed, a constant current will be fed to C1, and in consequence the voltage across it will rise in a linear fashion. In practice the
unit is adjusted so that PB1 needs to be closed for 10 seconds in order
for the voltage across C1 to reach the level that produces f.s.d. of the
meter.

If PB1 is only closed for 4 seconds, the meter will read 40% of f.s.d.,
if it is closed for 5 seconds, the meter will read 50% of f.s.d. The unit
thus acts as a simple stopclock.

When C2 is switched into circuit it will take ten times longer for f.s.d.
of the meter to be achieved, as C2 has ten times the capacity of C1.
The unit thus functions as a 0 – 10 sec. stopclock with C1 in circuit,
and a 0 – 100 sec. stopclock with C2 switched into circuit.

The values of C1 and C2 are such that these have to be electrolytic
types which have very wide tolerances (often as much as −20% and
+50%). It is therefore necessary to use a different preset resistor for
each range in the meter circuit so that the ranges can be individually
calibrated.

It is essential for the voltmeter section of the circuit to have an
extremely high input impedance as otherwise it would discharge the
timing capacitor when PB1 was released. The CA3130 with its virtually

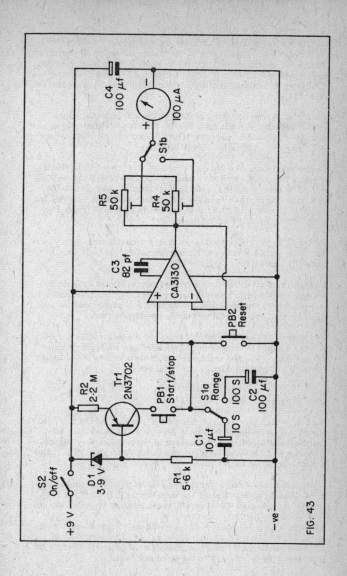

FIG. 43

infinite input impedance is ideal for this application. Provided the timing capacitors are good quality types, the meter reading should remain constant for at least a minute after PB1 is released. Any variation in the reading which does occur is almost certainly due to the deficiency of the timing capacitor rather than a lack of performance on the part of the CA3130.

The circuit is reset by operating PB2 which merely discharges the timing capacitor. It is advisable to push this two or three times as electrolytic capacitors can tend to regain some of their charge when only briefly short circuited a single time.

Calibrating the unit is quite simple, and all that is needed for this is an accurate timepiece which has a seconds hand. Initially adjust both R4 and R5 to insert maximum resistance into circuit. Then switch the unit on. Using the time piece as a reference standard, depress PB1 for precisely 10 seconds with S1 switched to the 10 second range. Then adjust R5 for exactly f.s.d. of the meter.

Switch S1 to the 100 second range, reset the circuit, and then depress PB1 for exactly 100 seconds. Then adjust R4 for precisely f.s.d. of the meter. The unit is then ready for use.

Note that one advantage of this type of stopclock over its more conventional counterpart is that it does not have to be activated manually. It is possible to use microswitches, acoustic and photo switches, and other types of switch to automatically operate the circuit.

The working life of PB2 might be increased if a resistor of a few ohms in value is added in series with it. This will limit the discharge current of the timing capacitors and prevent sparking occurring at the switch contacts (and also the damage this could cause).

Over Voltage Protection

An over voltage protection circuit is used to switch off the power to the circuit it is protecting if the supply rises above a safe level for this circuit. Op. amps. are ideal for use in this type of circuit when they are used as comparators. The circuit of such a device which employs a CA3130 I.C. is shown in Figure 44.

VR1 is adjusted so that at the required threshold voltage there is the same potential at both inputs. The inverting input is stabilised at a voltage of about 0.65V. by Tr1, which is used as an amplified diode.

When the supply voltage is at or below the threshold voltage, the output of the I.C. will be low, and Tr2 will be cut off. The relay will not therefore be energised, and its normally closed contacts will carry the positive supply rail through to the protected circuit.

If the supply rises above the threshold level, the voltage at the non-inverting input will go above that present at the inverting input. The output of the I.C. will go high, Tr1 will turn on and energise the relay, and the relays contacts will then break the supply line to the protected equipment.

In the event that the supply returns to a safe level again, the circuit will return to its original state and the supply will once again be connected to the protected equipment.

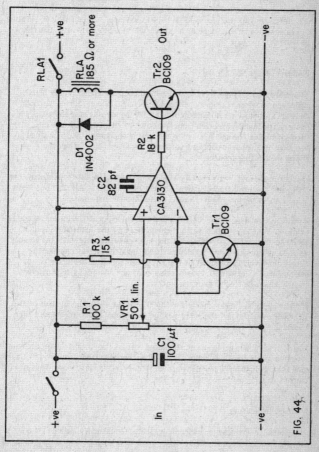

FIG. 44.

The circuit can be adjusted to operate at a threshold voltage of anything from 5 to 16 volts, which is the operating voltage range of the I.C. If it is possible that the input voltage could rise above 16V., this circuit is unsuitable for that application unless some form of voltage limiting is used to protect this circuit (which is not a very practical proposition). The relay operating voltage should be chosen to suit the range of voltages that are likely to be encountered when it is on. The voltages encounted when it is off are of no consequence in this respect.

Adjusting the circuit is quite simple. Apply a voltage to the circuit which is equal to the required threshold voltage. Adjust VR1 to produce the lowest possible voltage at its slider which does not cause the relay to turn off.

Under Voltage Shutdown

There are many types of circuit which incorporate a zener regulated supply, and where false operation will occur if the stabilised voltage should fall below its nominal voltage (due to an almost exhausted battery for instance). Such items of equipment often incorporate some form of under voltage shutdown, so that if the supply voltage does go too low, the apparatus ceases to function altogether, rather than giving misleading results.

It is a simple matter to modify the circuit of Figure 44 to perform this task, and the modified circuit is shown in Figure 45. All that has been done here is the inputs to the I.C. have been transposed.

As a result of this, the circuit operates in the opposite manner to the original. VR1 is adjusted so that the voltage at the inverting input is equal to the stabilised voltage at the non-inverting one with the supply voltage at the threshold level. Provided the voltage stays above the threshold level, the output of the I.C. stays low and the relay is not energised. The normally closed relay contacts conduct the positive supply through to the main circuitry.

If the supply voltage goes below the threshold level, the output of the I.C. will go high, and the relay will be energised. Its contacts will then break the positive supply line to the main equipment.

This circuit is set up in much the same way as the circuit of Figure 44. Apply a supply voltage which is equal to the required threshold voltage, and then adjust VR1 for the highest voltage at its slider that does not cause the relay to turn off.

Voltage Indicators

An alternative to having a circuit which automatically cuts off the supply when it goes above or below a certain limit, is to use one that will switch on a warning light to indicate an excessive or inadequate supply voltage.

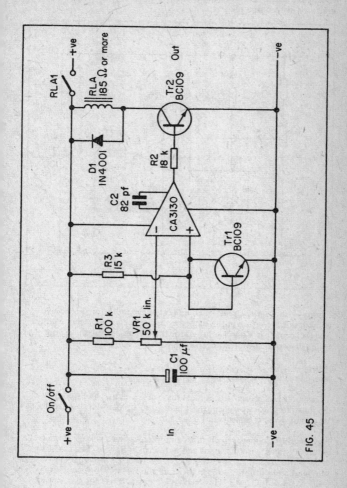

FIG. 45

The circuit diagram of an over voltage indicator is shown in Figure 46. This works in exactly the same manner as the circuit of Figure 44, except the output of the I.C. is used to operate a light emitting diode indicator instead of a relay circuit.

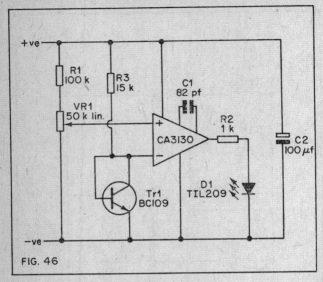

FIG. 46

Figure 47 shows the circuit of an under voltage indicator. These last two circuits are set up in the same way as their relay operating counterparts (except that VR1 is adjusted for the highest/lowest slider voltage which does not turn the indicator lamp off).

Morse Practise Oscillator

This simple audio oscillator is ideal for morse practise, and it also demonstrates how an op. amp. can be used in a phase shift oscillator. Its circuit diagram appears in Figure 48.

A phase shift oscillator requires an amplifier having a voltage gain of about 30 times, with the input and output in antiphase. The CA3130 is therefore used in the inverting mode. Feedback is provided between the input and output of the amplifier by the R-C network consisting of C4, R5, C3, R4, C2, R3. This feedback is not negative though, as each of the three R-C sections of this network provide a 60 degree phase shift at a certain frequency. They thus provide an overall 180 degree

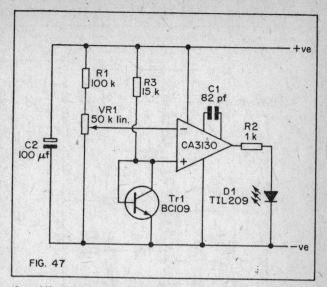

FIG. 47

phase shift, which brings the inverting input and the output into phase, and the circuit thus oscillates at the operating frequency of the phase shift network. With the specified values the circuit oscillates at approximately 1kHZ.

A phase shift oscillator provides a sinewave output, and it is therefore a good basis for a morse practise oscillator, as a morse signal received on a radio will also have a sinusoidal waveform. The unit can produce quite a realistic imitation of a real morse signal.

However, as was the case with the Wien bridge oscillators described earlier, the gain of the amplifier must be well controlled, as otherwise the circuit will either not oscillate, or will produce a distorted output. VR1 enables the circuit gain to be adjusted to the correct level.

The output of the oscillator is fed to VR2 which is the volume control. From here the signal is fed to a simple common emitter Class A output stage using Tr1. This drives a high impedance loudspeaker and a fairly high maximum volume can be obtained. Headphones of virtually any type and impedance can be used if preferred.

The morse key is simply inserted in series with the positive supply line so that the unit is normally off, and is turned on when the key is depressed. No separate on/off switch is required, since no current is consumed by the unit until the key is operated.

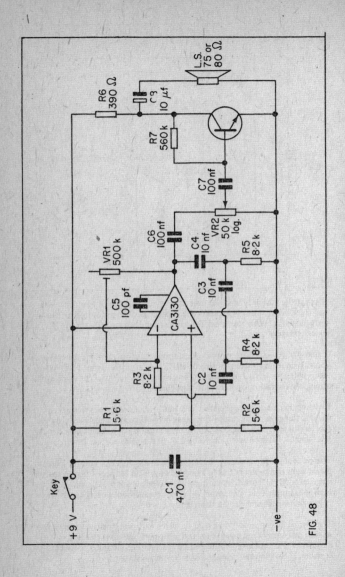

FIG. 48

Only one adjustment is necessary before the unit is ready for use, and that is to set VR1 at the correct point. Start with VR1 adjusted for maximum resistance and then switch the unit on. A rather rough note should be produced from the speaker, and adjusting VR1 for decreased resistance should improve the tone of the output. If it is taken too low in resistance, oscillation will cease.

Adjust VR1 to the point where it has the lowest resistance that produces reliable oscillation from the unit.

It is normally advisable to use 5% tolerance components in the phase shift network of this type of oscillator, but this circuit seems to work quite well using 10% or even 20% tolerance capacitors. It is merely then necessary to adjust VR1 for a higher gain in order to obtain reliable oscillation.

Electronic Heads Or Tails

This circuit shows how it is possible to electronically simulate the tossing of a coin. The circuit diagram of this device appears in Figure 49.

Tr1 and Tr2 form a conventional astable multivibrator which operates at a frequency of about 200HZ. This type of circuit produces antiphase squarewave outputs, the outputs being at the collectors of the transistors.

The CA3130 is used as a latch circuit, and this is basically identical in operation to the touch switch shown in Figure 41 of this book. Thus, when the supply is initially connected the output of the I.C. will go low, and the output load will receive no significant current. In this circuit the load for the I.C. is a light emitting diode indicator and its current limiting resistor (R8).

When PB1 is depressed, one output of the multivibrator is connected to the input of the latch. While PB1 is operated the lamp will be rapidly flashing on and off at a rate of about 200HZ. It comes on when the non-inverting input is taken above the potential at the inverting input by the input signal. The lamp is off when the input is in the low state.

Supply that PB1 is now released, and at the moments its contacts break the circuit the lamp is off. The lamp will remain off as the non-inverting input will be held in the low state as it is connected to the output via R7. If the output happened to be in the high state at the instant PB1's contacts broke the circuit, then it will remain in the high state with the lamp on. This is because the non-inverting input will now be held in the high state by being coupled to the output via R7.

It is a matter of chance whether the output happens to be in the high or low state at the moment and continuity through PB1 is broken, and there is no way of predicting this. It is not feasible to wait until the

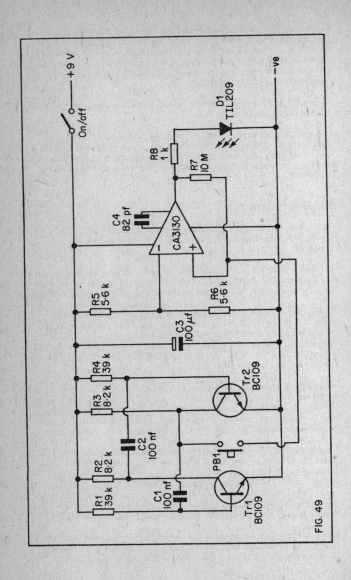

FIG. 49

lamp is either on or off and then release PB1, as the lamp is flashing at such a fast rate it appears to be on all the time that PB1 is depressed.

To simulate the tossing of a coin with this unit the high and low output states are called 'heads' and 'tails' respectively, or vice versa if preferred. The unit is switched on and PB1 is depressed briefly and then released. If the lamp is on after this it corresponds to 'heads', and if it is off this corresponds to a 'tails'.

The circuit will not provide precise 50/50 operation as the tolerances of the components will cause the multivibrator to generate a non-symmetrical squarewave. If, for instance, the output at Tr2 collector is high for longer than it is low, more 'heads' outputs will be obtained. If it is low for longer than it is high, there will be a 'tails' bias in the circuit.

Provided R1, R4, C1, and C2 are all close tolerance components (2% or better), any bias will be insignificant. Alternatively these can be ordinary 5 to 20% tolerance types, with a 100k preset resistor being used for R1. This preset is then adjusted to produce genuine 50/50 operation. This latter method is likely to be rather time consuming though, unless an oscilloscope is available. R1 can then be adjusted to produce an output from the multivibrator which has a 1 to 1 mark space ratio.

In the absence of a suitable oscilloscope it will be necessary to find the correct setting for R1 by trial and error. Take the unit through 50 or more operations each time and record the results. R1 is then adjusted slightly to correct any significant bias that becomes apparent. Eventually correct operation will be obtained.

Note that it is not possible for the I.C. to finish in an intermediate state, and the output will always go fully positive or negative when PB1 is released, even if this occurs while the multivibrator is changing state. If the output is fractionally above the level at the inverting input at the instant PB1 is opened, the non-inverting input will be taken above the level at the inverting input due to the presence of R7, and the output will go fully positive. If the output is slightly negative of the inverting input, then it will swing fully negative.

White Noise Generator

White noise is a sound that contains all audio frequencies, and an example of white noise is the background hiss that is present to some degree at the output of all amplifiers. White noise generators can be used as the basis of many sound effects when used in conjunction with envelope shapers and frequency filters. The circuit diagram of a white noise generator using a CA3130 is shown in Figure 50.

As mentioned above, all amplifiers generate noise, and the CA3130 is no exception to this rule. Normally the noise level it produces is quite small as it is reduced to an insignificant level by the use of negative feedback. In this circuit the I.C. is used with a high voltage gain so that there is less feedback and more noise. The level of noise at the output is still rather low, and a high gain common emitter amplifier is used to boost this to a reasonable level. This stage is built around Tr1.

The unit has a wideband output of about 250mV. R.M.S.

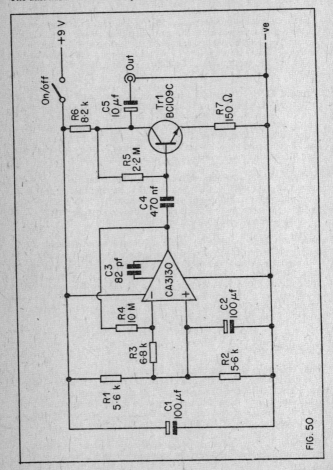

FIG. 50

The circuit has a very high overall gain, and a sensible layout should be used in order to avoid instability due to stray feedback.

Current Limiter

Most bench power supplies seem to incorporate a current limiting circuit these days. This is more convenient than using a fuse, as if an overload occurs the limiter circuit prevents the output current from reaching an unsafe level. When the overload is removed from the output of the P.S.U., the unit functions quite normally. Thus the expense and bother of replacing fuses is eliminated

This type of circuit also has another important advantage over using a conventional fuse. The speed at which even a fast blow fuse operates is low in comparison to an electronic cutout or limiter. Semiconductor devices can be destroyed by very brief current surges, and so an electronic protection circuit provides a worthwhile increase in protection efficiency.

The circuit diagram of a current limiter using a CA3130 I.C. is shown in Figure 51. R1 and Tr1 form a voltage stabilising circuit with an output of approximately 0.65V. R2 and R3 are a simple potential divider which are fed from this regulated voltage, and the output of this network is fed to the non-inverting input of the I.C. About 100mV. is fed to this input.

When low output currents are drawn from the circuit only a very low voltage is developed across RX, and the inverting input is below the potential at the non-inverting input. The output of the I.C. is therefore in the high state, with Tr2 turned hard on.

If the output current is steadily increased, a point will be reached where the voltage across RX is more than the voltage at the non-inverting input. This will cause the output of the I.C. to swing slightly negative in order to turn Tr2 off to some degree, and so reduce the output current. Even if the output is short circuited, the output current will not rise above the threshold level. The negative feedback action of the circuit will always result in the voltage across RX being limited to no more than that present at the non-inverting input. As the voltage across RX is limited to a certain level, the output current that flows through it must also be limited.

This circuit has a couple of advantages over the more usual ones that are used. Firstly, in any current limiting circuit there is a voltage drop between the input and output, even on outputs which are below the threshold current. This is normally a drop of something like 0.65V. In this circuit the voltage drop is only typically something in the order of 0.15V. This means that less heat is generated in the series resistor (RX) and a greater maximum output voltage is available.

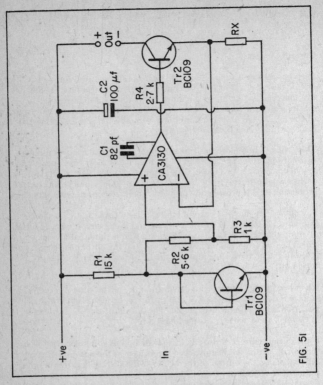

FIG. 51

The second advantage is that with this circuit there is not output voltage drop when the unit is passing output currents a little below the threshold level. The output current threshold is so well defined that it is not really possible to significantly improve upon the circuit in this respect.

No value is shown for RX, and the value of this must be chosen to suit the particular output threshold current required. The output current reaches its maximum level when 100mV. is developed across RX. Thus if it has a value of 10 ohms the output current will be 10mA., if it is 100 ohms the output threshold will be 1mA. The equation for calculating the value of RX is:—

RX = 0.1 divided by the threshold current in amps.

Although this unit has been described as for use in bench power supplies, it can also be used in other equipment, such as in the supply lines of an audio amplifier so that it provides output overload protection.

High Current Version

The unit as described above is only suitable for use at relatively modest output currents. The circuit can operate at any input voltage of between 5 and 16 volts, and the maximum output current that can be safely handled is dependant upon the input voltage. This is about 20mA. at 16 volts and about 60mA. at 5 volts.

It is a simple matter to modify the circuit to operate safely at higher currents, and all that is required is to change Tr2 to a 2N3504 power transistor, and reduce R4 to 220 ohms. The unit can then operate properly at currents up to about 500mA. regardless of the input voltage, but Tr2 must be given adequate heatsinking.

Over Current Indicator

Another variation on the current limiter is the over current indicator circuit which appears in Figure 52. This operates in much the same way as the original circuit with the output of the I.C. normally high. This causes the L.E.D. indicator in the output circuit of the I.C. to be off.

If an excessive current is drawn from the unit the voltage at the inverting input will go above that at the non-inverting input. The output of the I.C. will then swing negative and D1 will come on, indicating that an excessive output current is being drawn. When the overload is removed the circuit returns to normal with the L.E.D. switching off.

The value of RX is calculated in exactly the same way as it was for the previous circuit.

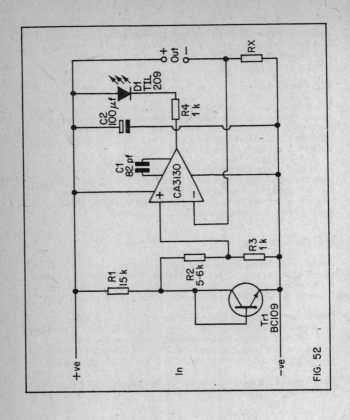

FIG. 52

COMPONENTS

In many of the circuits described in this book the CA3130 is not the only semiconductor device which is employed. Figure 53 shows the leadout diagrams for the other semiconductors used in the circuits in this book, including the diodes and rectifiers.

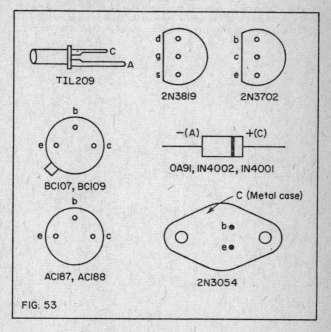

FIG: 53

There are many alternatives to most of these additional semiconductors, but in general the types specified are less expensive and more readily available than these alternatives. There is therefore little point in using devices other than those specified unless a suitable component is already to hand.

Unless otherwise noted, all the resistors are ordinary miniature 1/4 or 1/3 watt types having the usual tolerances (5% up to 1 Meg., and 10% above 1 Meg.). All potentiometers are carbon types and the circuit diagrams specify whether a log. or a lin. type is required.

All capacitors must be of good quality, and ceramic plate, silvered mica, or polystyrene types can be used for values up to 10nf. For 10nf and above, apart from electrolytic types, any good quality plastic foil (polyester, polycarbonate, etc.) components can be used.

Electrolytic capacitors can be any good quality type, and if in doubt about a suitable voltage rating for any of these, use one having a working voltage greater than the supply potential used.

A couple of circuits use trimmer capacitors (the 'M.W. Radio' and the 'Crystal Calibrator'), and at the frequencies involved in these virtually any trimmer of around the specified value should perform satisfactorily.

Notes

Notes

Notes

Notes

Notes

Notes

Notes

Notes

Please note overleaf is a list of other titles that are available in our range of Radio and Electronic Books.

These should be available from all good Booksellers, Radio Component Dealers and Mail Order Companies.

However, should you experience difficulty in obtaining any title in your area, then please write directly to the publisher enclosing payment to cover the cost of the book plus adequate postage.

BABANI PRESS & BERNARDS (PUBLISHERS) LTD
THE GRAMPIANS
SHEPHERDS BUSH ROAD
LONDON W6 7NF
ENGLAND

BP1	First Book of Transistor Equivalents and Substitutes	60p
BP2	Handbook of Radio, TV and Ind. & Transmitting Tube & Valve Equiv.	60p
BP6	Engineers and Machinists Reference Tables	50p
BP7	Radio and Electronic Colour Codes and Data Chart	25p
BP11	Practical Transistor Novelty Circuits	40p
BP14	Second Book of Transistor Equivalents	1.10p
BP22	79 Electronic Novelty Circuits	1.00p
BP23	First Book of Practical Electronic Projects	75p
BP24	52 Projects using IC741	95p
BP26	Radio Antenna Handbook for Long Distance Reception & Transmission	1.00p
BP27	Giant Chart of Radio Electronic Semiconductor & Logic Symbols	60p
BP28	Resistor Selection Handbook (International Edition)	60p
BP29	Major Solid State Audio Hi-Fi Construction Projects	85p
BP30	Two Transistor Electronic Projects	85p
BP31	Practical Electrical Re-wiring & Repairs	85p
BP32	How to Build Your Own Metal and Treasure Locators	1.00p
BP33	Electronic Calculator Users Handbook	95p
BP34	Practical Repair & Renovation of Colour TV's	1.25p
BP35	Handbook of IC Audio Preamplifier & Power Amplifier Construction	1.25p
BP36	50 Circuits Using Germanium, Silicon and Zener Diodes	75p
BP37	50 Projects Using Relays, SCR's and TRIAC's	1.25p
BP38	Fun & Games with your Electronic Calculator	75p
BP39	50 (FET) Field Effect Transistor Projects	1.25p
BP40	Digital IC Equivalents and Pin Connections	2.50p
BP41	Linear IC Equivalents and Pin Connections	2.75p
BP42	50 Simple L.E.D. Circuits	95p
BP43	How to make Walkie-Talkies	1.25p
BP44	IC 555 Projects	1.75p
BP45	Projects on Opto-Electronics	1.25p
BP46	Radio Circuits using IC's	1.35p
BP47	Mobile Discotheque Handbook	1.35p
BP48	Electronic Projects for Beginners	1.35p
BP49	Popular Electronic Projects	1.45p
BP50	IC LM3900 Projects	1.35p
BP51	Electronic Music and Creative Tape Recording	1.25p
BP52	Long Distance Television Reception (TV-DX) for the Enthusiast	1.45p
BP53	Practial Electronic Calculations and Formulae	2.25p
BP54	Your Electronic Calculator and Your Money	1.35p
BP55	Radio Stations Guide	1.45p
BP56	Electronic Security Devices	1.45p
BP57	How to Build your own Solid State Oscilloscope	1.50p
BP58	50 Circuits using 7400 Series IC's	1.35p
BP59	Second Book of CMOS IC Projects	1.50p
BP60	Practical Construction of Pre-Amps, Tone Controls, Filters and Alternators	1.45p
BP61	Beginners Guide to Digital Techniques	95p
BP62	Elements of Electronics – Book 1	2.25p
BP63	Elements of Electronics – Book 2	2.25p
BP64	Elements of Electronics – Book 3	2.25p
BP65	Single IC Projects	1.50p
BP66	Beginners Guide to Microprocessors and Computing	1.75p
BP67	Counter Driver and Numeral Display Projects	1.75p
BP68	Choosing and Using Your Hi-Fi	1.65p
BP69	Electronic Games	1.75p
BP70	Transistor Radio Fault-Finding Chart	50p
BP71	Electronic Household Projects	1.75p
BP72	A Microprocessor Primer	1.75p
BP73	Remote Control Projects	1.95p
126	Boys Book of Crystal Sets	40p
160	Coil Design and Construction Manual	75p
196	AF-RF Reactors – Frequency Chart for Constructors	15p
202	Handbook of Integrated Circuits (IC's) Equivalents and Substitutes	1.00p
203	IC's and Transistor Gadgets Construction Handbook	60p
205	First Book of Hi-Fi Loudspeaker Enclosures	95p
207	Practical Electronic Science Projects	75p
208	Practical Stereo and Quadrophony Handbook	75p
210	The Complete Car Radio Manual	1.00p
211	First Book of Diode Characteristics Equivalents and Substitutes	1.25p
213	Electronic Circuits for Model Railways	1.00p
214	Audio Enthusiasts Handbook	85p
215	Shortwave Circuits and Gear for Experimenters and Radio Hams	85p
218	Build Your Own Electronic Experimenters Laboratory	85p
219	Solid State Novelty Projects	85p
220	Build Your Own Solid State Hi-Fi and Audio Accessories	85p
221	28 Tested Transistor Projects	1.25p
222	Solid State Short Wave Receivers for Beginners	1.25p
223	50 Projects using IC CA3103	1.25p
224	50 CMOS IC Projects	95p
225	A Practical Introduction to Digital IC's	95p
226	How to Build Advanced Short Wave Receivers	1.20p
227	Beginners Guide to Building Electronic Projects	1.25p
228	Essential Theory for the Electronics Hobbyist	1.25p
RCC	Resistor Colour Code Guide	20p